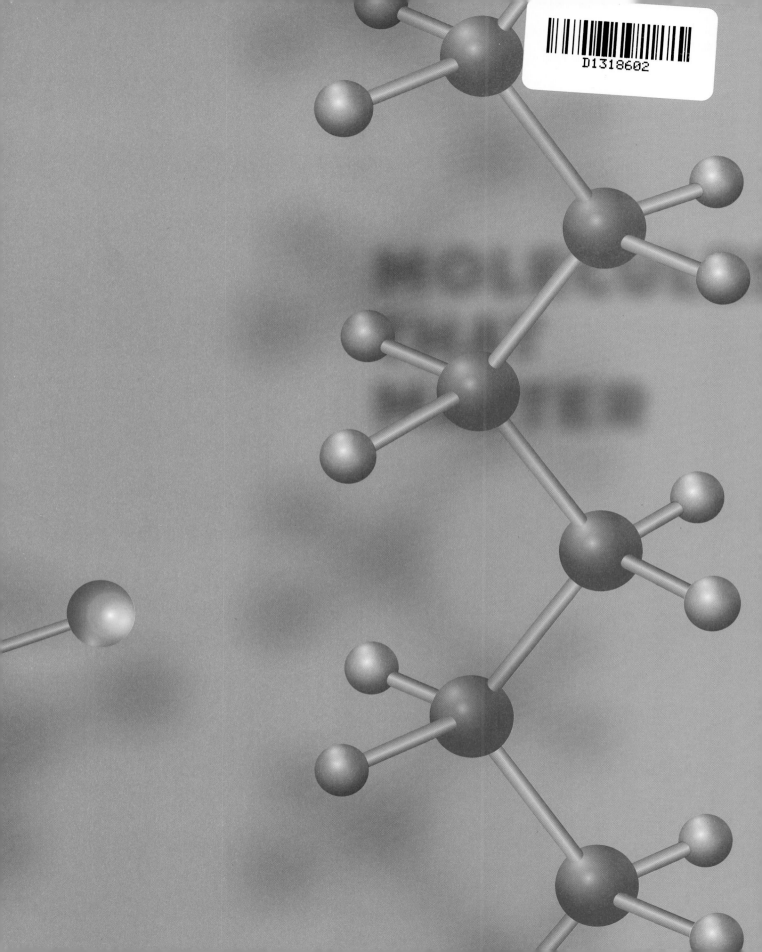

MOLECULES THAT MATTER

MOLECULES

THAT MATTER

Edited by Raymond J. Giguere

**The Frances Young Tang
Teaching Museum and Art Gallery
at Skidmore College**
Saratoga Springs, New York

Chemical Heritage Foundation
Philadelphia, Pennsylvania

The Frances Young Tang Teaching Museum
and Art Gallery at Skidmore College
www.skidmore.edu/tang

Chemical Heritage Foundation
www.chemheritage.org

Printed and bound in Italy

Library of Congress
Cataloging-in-Publication Data
Molecules that matter / edited by
Raymond J. Giguere.
   p. cm.
   ISBN 0-9765723-7-0
   1. Molecules—Exhibitions. 2. Matter—
Constitution—Exhibitions. 3. Molecules—
Research—Social aspects—Exhibitions.
4. Scientific discoveries—History—20th
century—Exhibitions. I. Giguere, Raymond
J. II. Chemical Heritage Foundation.
III. Frances Young Tang Teaching Museum
and Art Gallery. IV. Title.
   QC173.M646 2008
   540.74'74748—dc22   2008033780

CONTENTS

Penicillin and nylon
Installation view, Tang Museum

WHY MOLECULES MATTER

The story of the chemical and molecular sciences is central to the story of human achievement in the twentieth century, and developments in these sciences will continue to shape the progress and promise of our new century. Chemists and chemical engineers constantly improve our material world with better communications, colors, fabrics, fragrances, furnishings, pharmaceuticals, and food production. But material achievements wrought by the chemical and molecular sciences are quickly taken for granted: we expect effective medicines to fight maladies and diseases; we expect a wide range of food choices regardless of the seasonality of crops; we expect the ability to interface instantaneously over great distances in our work and personal lives. As the chemical and molecular sciences continue to meet and exceed these expectations, we can only imagine how their story will develop.

The idea for this exhibit—identifying a molecule for each decade of the twentieth century and exploring its scientific, historical, and sociological dimensions—was the brainchild of Raymond J. Giguere, Class of 1962 Term Professor of Chemistry at Skidmore College. Early in 2004, Giguere and his colleagues at the Frances Young Tang Teaching Museum and Art Gallery visited the Chemical Heritage Foundation (CHF) to propose a collaboration that would advance the educational missions of both organizations. Starting in 2005, a joint committee was formed, and the Scientific Advisory Board Giguere had convened narrowed the choice to the final ten organic compounds—a selection that subsequently was reviewed by Nobel laureates Roald Hoffmann and Dudley Herschbach, at the invitation of CHF.

After the molecules were identified, a team of Skidmore and CHF colleagues undertook additional research, commissioning and identifying art for loan, tracking down illustrative artifacts, and working with fabricators to create the molecular models. Co-curated by Giguere and John S. Weber, Dayton Director of the Tang Museum, *Molecules That Matter* reveals the academic and industrial research that led to discovery and dissemination of these substances, illuminates their social consequences, and links them to the lives of museum visitors in ways too numerous, unexpected, and thought-provoking to catalog here.

The tradition of liberal education has long challenged us not to lead unexamined lives. Skidmore's Tang Museum provides a unique "space" for exploring the contemporary implications of this principle. *Molecules That Matter* extends this examination by bringing to the fore the linkage of science (and technology) and the history of the twentieth century. As the exhibit demonstrates, the chosen molecules inform huge segments of modern life. Yet most are poorly understood and certainly are easily overlooked because of our familiarity with them. Indeed, the influence of science in general (and, perhaps, organic chemistry in particular) is often invisible within our quotidian experience. Yet without science and technology, we would not be who we think we are, and our lives would proceed along paths far different from those we routinely follow. For those lacking in-depth knowledge of organic chemistry (and organic synthesis) or the history of science and technology, the information highlighted in this exhibit is likely to come as a revelation. And that is precisely the point. Ultimately, it provokes us to reassess our understanding of both our *selves* and our *world* in light of the pervasive influence of these ten molecules that matter.

Just as *Molecules That Matter* blurs the boundaries of art, natural science, social science, and other disciplines, it encourages visitors to become better informed and more highly engaged citizens of our rapidly changing, often troubled world. This world demands of each of us a capacity to understand the nature of modern scientific argument; the ability to weigh competing sets of data, assertions of value, and risk; and the willingness to relate all of these to questions of public policy—questions that often concern a daunting set of socio-scientific issues. Genetically modified organisms, global warming, stem-cell research, federal drug policies, medical litigation and health-care rights and costs, oil dependence and global politics: all these issues and more are implicated in *Molecules That Matter*! As such, the exhibition is not only a treasure trove of ideas and teachable controversies for undergraduate learning but also a fascinating stimulus for us all to think further about the fundamental role of science and technology in contemporary life.

We thank all the contributors at Skidmore College and at CHF for participating in this exhibit. We hope that *Molecules That Matter* inspires each of us to see how the ongoing enterprise of the chemical and molecular sciences is deeply rooted in human experience, human need, and ultimately in human nature.

PHILIP A. GLOTZBACH
PRESIDENT, SKIDMORE COLLEGE

ARNOLD THACKRAY
CHANCELLOR, CHEMICAL HERITAGE FOUNDATION

Polyethylene, progestin, and aspirin
Installation view, Tang Museum

PREFACE

I work this wild chemical garden with one old tool.
Let me show others new ways to see.[1]

—ROALD HOFFMANN, NOBEL LAUREATE, 1981

Molecules That Matter explores ten organic molecules—aspirin (1900), isooctane (1910), penicillin G (1920), polyethylene (1930), nylon 6,6 (1940), DNA (1950), progestin (1960), DDT (1970), Prozac (1980), and buckyball and carbon nanotubes (1990)—that have shaped the course of humanity throughout the twentieth century. While any number of other molecules could have been chosen for this exhibit, the impact of these ten is nothing short of astonishing.

If we ask how our lives differ from those of our parents or grandparents, it becomes quite easy to see the effect molecular chemistry has had on successive generations in the last century. My parents, both born in 1913, could obtain aspirin, although that first wonder drug was just coming onto the market, but they had no access to reliable and safe methods of birth control and could readily have died in childhood from bacterial infections from which I was protected by penicillin and modern antibiotics. Their families had no

•

Frank Moore
Beacon, 2001
Installation view, Tang Museum

cars; traveling thirty miles to the nearest town to see a doctor, my dad would say, was like "going to the ends of the earth." Times really have changed. So have we.

In the twentieth century our knowledge of substances at the molecular level has significantly redefined our world—even life itself. How we have changed and who we have become as a result of this remarkable molecular revolution is the overarching story this exhibition conveys. *Molecules That Matter* tells this story by examining the influence of just ten molecules, one associated with each decade of the past century. Most of us interact with these molecules often, perhaps even daily, and though few of us know their scientific names, origins, or development, their significance cannot be denied.

The task of selecting the ten *Molecules That Matter* finalists from hundreds of potential molecular candidates fell to a dedicated and talented volunteer Scientific Advisory Board of professionals drawn from academia, industry, and the Chemical Heritage Foundation. From the onset it was clear that the selection process would take time, patience, and, above all, iteration. It was also clear that no matter which set of ten we chose, no ten substances could adequately represent the complexity of twentieth-century chemical history. After several rounds of e-mail discussions covering many excellent possibilities, we winnowed the candidates down to twenty. Among these were favorites like aspirin, penicillin, and DNA—all unanimously perceived as the best choices for their respective decades. Conversely, selecting other finalists was not so easy. The 1980s, for example, ultimately pitted taxol and Prozac against each other. Taxol, the effective and powerful breast and ovarian cancer drug, would have permitted the exhibit access to this major disease. Prozac, on the other hand, enjoys wider name recognition and would allow the exhibition to address the treatment of mental illness. The vote was close: Prozac prevailed by a slim margin. Runner-up candidates for other decades were rayon, cortisone, styrene, and cholesterol.

Careful consideration was also given to finding a proper balance among the various categories of organic molecules represented in the exhibit. *Molecules That Matter* could conceivably have focused exclusively on ten polymers, for example, or ten medicinals, or ten natural products, and so on. In the end, we selected four medicinals, three natural products, two synthetic polymers, and two "others"—a key hydrocarbon related to fuels and the world's best-known pesticide. Another board would have generated a somewhat different list of molecular finalists, but there is no doubt that each of the *Molecules That Matter* substances is a bona-fide titan in its historical effect.

Although unimaginably small, molecules paradoxically have immense influence—hence the super-sized, beautifully crafted molecular models fabricated especially for the exhibition. Based on the symbolic ball-and-stick models and drawings scientists use to express the identity of molecules, the models remind us of the enormous significance of this infinitesimal world and reveal its language. Complex but elegant, this language is common to those trained in the discipline and, with brief introduction, accessible to those outside it. These models and the substances they represent truly are, in their impact, "bigger than we are."

Our ability to understand the structure and language of the molecular world and use that knowledge to improve our lives is arguably the most powerful tool for positive change that humankind has yet developed. Molecules, formed when two or more atoms bond to each other, are unique, discrete entities. They are created only when precise orders of atomic connections are achieved, synthesized either by nature or by human hand. In this manner, molecules possess a gestalt quality: removing just one atom, or relocating even one molecular bond, compromises a molecule's identity, consequently completely affecting what substance it is and how it interacts in the world.

Along with molecular models, *Molecules That Matter* features contemporary art by nationally and internationally recognized artists and a range of objects and historical documents drawn from popular culture, advertising, and publishing. The artworks in the exhibition function as a mirror and index of how thoroughly these molecules have penetrated our consciousness, echoing the presence of substances like aspirin, nylon, and polyethylene in our lives. Artists use them as the raw material for their art and as an inspiration for their explorations. Always we chose artists and works that would inject new issues and ideas into the exhibition, opening additional avenues of speculation and encouraging visitors to see connections between hard science and their daily lives.

Our selection of documents, cultural objects, historical artifacts, books, and ephemera was similarly open-ended and, at times, deliberately playful. We selected a varied range of period advertisements, books, magazines, toys, manufactured items, and other objects that suggest the dimensions and texture of larger stories we can only indicate within the framework of the exhibition itself. The impact of each molecule in the exhibition is so pervasive and multifaceted that we can only hint at its full scale.

Finally, the foundation for our molecular understanding has come through the brilliant, often dogged work of dedicated professional scientists since the humble beginnings of organic chemistry around 1850; thus, *Molecules That Matter* presents the human struggles and compelling stories of the talented women and men who labored to unlock the secrets of this world. These individuals and teams serve as guides and translators, helping us learn the distinctive molecular language and appreciate its meaning, potential, and power.

The goal of *Molecules That Matter* is to open visitors' minds to the role of molecular chemistry in all our lives, to provoke new insights, and to make the invisible world of molecules visible. We offer the exhibition, catalog, and Web feature as a beginning to a process of discovery, exploration, and enthusiasm for chemistry that will, we hope, extend well beyond the boundaries of this project and into the future. Long after these words are printed, the ten molecules in this exhibition, and many like them, will be cultivated for their inherent value and passed forward into the collective march of human history like priceless family heirlooms.

RAYMOND J. GIGUERE
CLASS OF 1962 TERM PROFESSOR, CHEMISTRY DEPARTMENT, SKIDMORE COLLEGE

1. Roald Hoffmann, "Somewhere," in *Gaps and Verges* (Gainesville: University Press of Florida, 1990), 72.

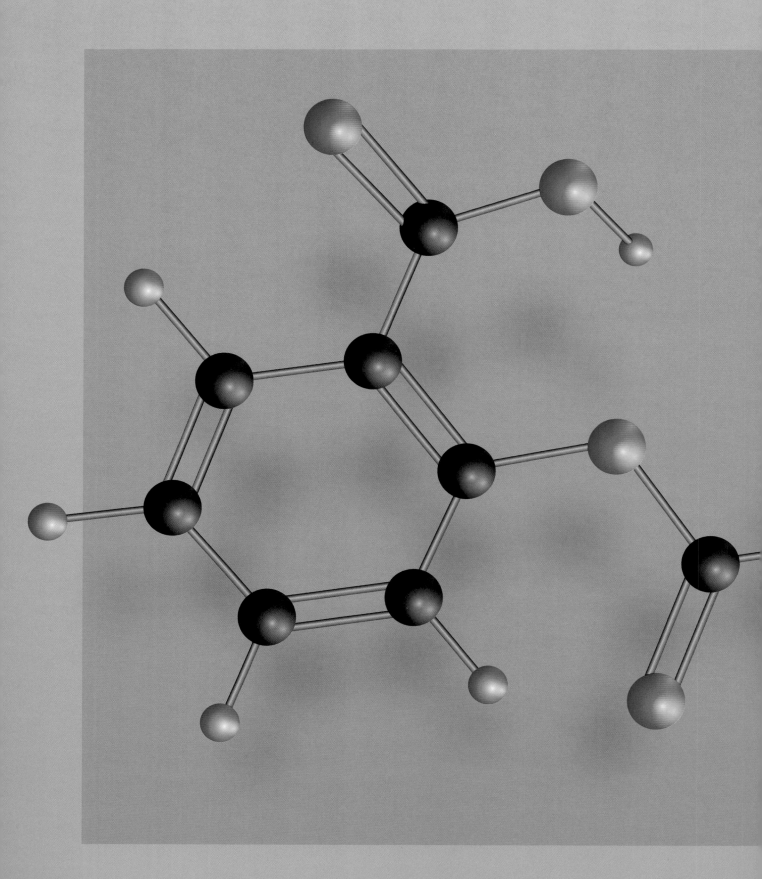

1900

ASPIRIN

$C_9H_8O_4$

GERMAN PATENT NUMBER: 671,769
CAS REGISTRY NUMBER: 50-78-2

Headache? Fever? Muscle pain? "Take two aspirin and call me in the morning." Like most of us, when you experience everyday aches and pains, a bottle of aspirin is probably the first thing you reach for. Yet, while aspirin has been one of the most popular pharmaceutical agents of the past one hundred years, it is actually a synthetic derivative of the natural substance salicylic acid, the associated healing properties of which have been known for millennia.[1]

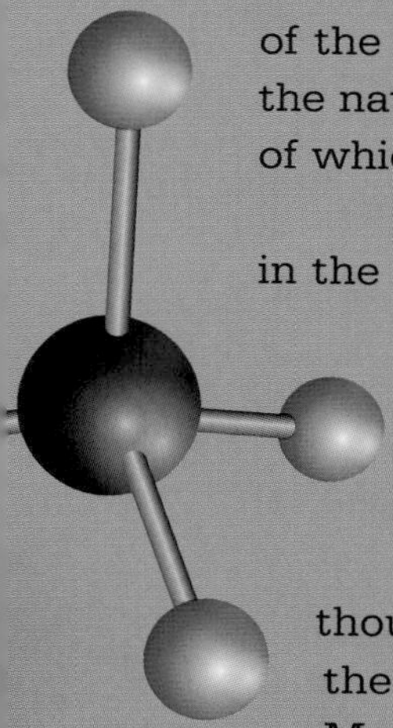

Salicylic acid is a main component of an herbal extract found in the bark of a number of trees, including the willow tree, and in a number of fruits, grains, and vegetables.[2] As such, salicylic acid—and related salicylates—have long been common components of a normal human diet, functioning as a natural defense against what we consider common ailments today.[3]

The first recorded use of salicylates dates back some four thousand years to the Sumerians, who noted the pain remedies of the willow tree on early clay tablets.[4] Ancient civilizations in Mesopotamia used the extract from willow trees to treat fever,

by Daniel R. Goldberg

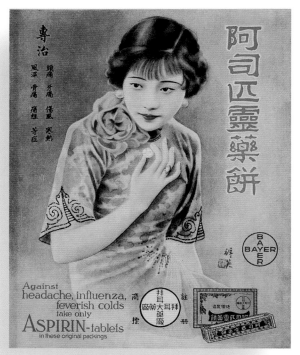

pain, and inflammation. Both Chinese and Greek civilizations employed willow bark for medical use more than two thousand years ago, and the Chinese also used poplar bark and willow shoots to treat rheumatic fever, colds, hemorrhages, and goiter.[5] One of the most noteworthy reports of the use of salicylic acid comes from the father of modern medicine, Hippocrates (460–370 BCE), who recommended chewing on willow-tree bark for patients suffering from fever and pain, as well as the use of a tea brewed from willow bark given to women to lessen pain during childbirth.[6] Around 100 CE the Greek physician Dioscorides prescribed willow bark as an anti-inflammatory agent.[7]

Despite this long history, it was not until 1763 that the Reverend Edward Stone of the Royal Society of London conducted one of the first clinical studies on the effects of willow-bark powder by treating patients suffering from ague (a fever thought to be caused by malaria).[8] And approximately one hundred years later the Scottish physician Thomas MacLagan studied the effects of willow powder on patients suffering from acute rheumatism, demonstrating that it could relieve fever and joint inflammation.[9]

The chemical investigation of the healing properties of the substance within the willow bark had already begun in earnest, however,

ABOVE Bayer aspirin advertising poster, c. 1950
OPPOSITE Bayer aspirin advertising poster, 1952

during the early nineteenth century. This investigation was driven in part by Napoleon's continental blockade on imports, which affected suppliers of Peruvian cinchona-tree bark (another natural source of salicylic acid).[10] In 1828 Johann Büchner, a professor at the University of Munich, isolated a yellow substance from the tannins of willow trees that he named *salicin,* the Latin word for *willow.* A pure crystalline form of salicin was isolated in 1829 by Henri Leroux, a French pharmacist, who then used it to treat rheumatism. In the late 1800s large-scale production of salicylic acid for the treatment of pain and fever was initiated by the Heyden Chemical Company in Germany.[11]

The beginning of aspirin as we know it today dates from the same period, when Friedrich Bayer and Company, a former dye-manufacturing firm in Germany, began to shift its focus from the diminishing dye industry to pharmaceutical production. Because the Bayer Company was already well known, it easily developed brand-name recognition as a pharmaceutical maker. The company's shift to pharmaceuticals production coincided serendipitously with a boom in new pharmaceutical agents, making it seem that a new drug was put on the market almost daily.[12]

Just as the medical benefits of salicylic acid had long been known, so too had some of the health issues related to prolonged use of large doses of the drug. Such use often led to gastrointestinal irritation, which could in turn lead to nausea, vomiting, bleeding, and ulcers. In 1895, to counteract such problems, the head of chemical research at Bayer, Arthur Eichengrün, assigned the task of developing a "better" salicylic acid to one of the company's chemists, Felix Hoffmann. Eventually cited by many as the discoverer of aspirin, Hoffmann

Aspirin is believed to be the most widely used pharmaceutical in the world; it is the main ingredient in more than fifty over-the-counter drugs.

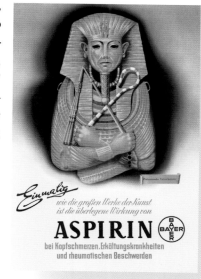

approached the task with a personal interest: his father suffered from rheumatism and was taking salicylic acid for it, but he could no longer ingest the drug without vomiting.[13] Hoffmann's search of the scientific literature yielded a way to alter salicylic acid chemically through modification of the hydroxyl group on the benzene ring. The key to his discovery, although realized only later, was that this chemical transformation provided a new molecule that the body could absorb without significant gastrointestinal distress. Once ingested, the new molecule was converted back to salicylic acid in the stomach, liver, and blood, thereby providing the desired therapeutic benefits. As such, modern synthetic aspirin can be considered a drug-delivery system for a natural product that has been in medical use for literally thousands of years.

However, this new derivative of salicylic acid did generate some controversy. There was a difference in opinion regarding the potential benefits of acetylsalicylic acid, which would ultimately become a personal dispute as well as a scientific one. Heinrich Dreser, who was responsible for the standardized testing of pharmaceutical agents, disagreed with Eichengrün's approach to the drug. Eichengrün had distributed Hoffmann's compound to local physicians, whereas Dreser had no initial interest in supporting the new drug. Ironically, it would be Dreser who published the first article on aspirin, probably because his contract with Bayer provided him royalties for any drug he introduced; Hoffmann and Eichengrün could only gain monetary reward on patentable compounds. In the article Dreser compared aspirin with other salicylates in an effort to demonstrate that it was more beneficial and less toxic. This work was coupled with human trials whose results were published in 1899 in the journals *Die Heilkunde* and *Therapeutische Monatshefte*, showing that aspirin

Only two weeks after Hoffmann synthesized aspirin, he also synthesized diacetylmorphine, commonly known as heroin.

Fred Tomaselli
13,000, 1996
Aspirin, acrylic, and resin on wood panel
48 x 48 inches

Aspirin
Installation view, Tang Museum

was indeed superior to other known salicylates.[14] On March 6, 1899, the Bayer Company registered the product under the trade name Aspirin and then actively began to distribute the white powder to hospitals and clinics.[15]

It costs about 1.5 cents to produce one aspirin pill.

There are two theories concerning the origin of the name *aspirin*. According to the first, the word derives from St. Aspirinius, a Neapolitan bishop who was the patron saint of headaches. According to the second theory, the name comes from the combination of *acetyl;* the Latin *Spiraea*, the genus of plants to which meadowsweet belongs and which also contains salicylic aldehyde, a precursor to salicylic acid (in German *salicylic acid* is *Spirsäure*); and *–in,* which was a common ending for drug names at the time.[16] Although the company name Bayer has long been associated with aspirin, after World War I, Bayer lost the sole right to use the name *aspirin.* It was acquired in 1919 by Sterling Incorporated for the then unheard-of price of $3 million, along with Bayer's U.S. drug properties.[17] Eventually Bayer reacquired the trademark from SmithKline Beecham as part of a wider deal, for the price of $1 billion.[18]

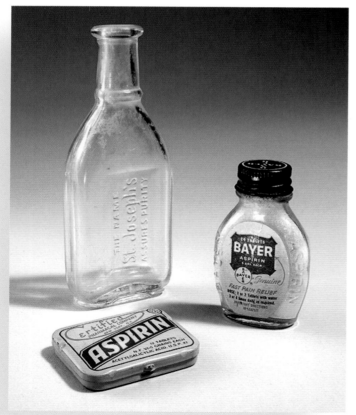

The first tablet form of aspirin appeared in 1900, creating an ease of use that quickly expanded the drug's recognition among professionals. Medical reports highlighted the benefits of aspirin, and its popularity reflected the already significant use of salicylic compounds, coupled with the fact that this new drug was considerably safer and comparably less toxic. In 1915 aspirin became available to the public without a prescription, making it arguably the first modern, synthetic, over-the-counter, mass-market medicine and a household name around the world.

By providing an easy and inexpensive method to alleviate pain, aspirin began to change the experience and expectations of patients and doctors and ultimately the nature of modern medicine itself. Before the mid-1800s Western physicians had considered pain an essential diagnostic tool, something that aspirin alleviated and thus disguised.[19] Doctors would now have to look to other symptoms.

It was not until 1971 that scientists began to understand how aspirin worked in the body as an anti-inflammatory agent—what is now referred to as a nonsteroidal anti-inflammatory drug (NSAID). John Robert Vane, a British pharmacologist, and his graduate student Priscilla Piper performed pioneering work on aspirin, exploring the effects of the drug on isolated lungs from guinea pigs and studying the effects of substances released from the lungs during severe allergic reactions to aspirin.[20] During these studies the scientists identified two uncharacterized substances, one of which turned out to be a prostaglandin, a hormone-like compound involved in causing diverse effects in the body, including vasodilation, vasocontraction, and sending messages of pain and discomfort to the brain. Piper and Vane later discovered that this prostaglandin had an effect similar to a known enzyme responsible for the contraction of nonvascular smooth muscle. Further studies demonstrated that aspirin minimized some effects of vasodilation response, ultimately leading Vane to consider that aspirin was inhibiting the synthesis of prostaglandins.[21] For Vane's pioneering work he, along with Sune K. Bergström and Bengt I. Samuelsson, received the Nobel Prize in Physiology or Medicine in 1982.

In the 1930s Bayer was part of I.G. Farben, a German industrial conglomerate that helped bankroll Adolf Hitler and the Nazi Party. After World War II, a number of I.G. Farben executives were convicted of war crimes, including the use of slave labor in I.G. Farben factories located near concentration camps.

OPPOSITE Bayer and St. Joseph's aspirin containers, c. 1900–1950
ABOVE Bayer aspirin advertising sign, c. 1920

Aspirin
Installation view, Tang Museum

But how does aspirin affect the production of prostaglandins? In 1976 researchers discovered a particular enzyme, cyclooxygenase, or COX, that is responsible for producing a number of biological mediators, including prostaglandins. Aspirin was found to bind selectively and irreversibly to this enzyme, providing the drug's beneficial properties. This characteristic differs from that of other well-known NSAIDs (e.g., ibuprofen), which are reversible inhibitors.[22] Further research indicated that there was not one COX enzyme, but three, and that each played a different role in the human body.[23] While one COX enzyme is responsible for the synthesis of prostaglandins during inflammatory reactions, a second is involved in producing prostaglandins that help protect the stomach mucosa. Aspirin affects both enzymes, providing analgesic effects as described, but at high doses, resulting at times in stomach irritation. In an effort to separate the two effects pharmaceutical companies have worked hard to develop selective COX inhibitors, such as Celebrex, Vioxx, and Mobic, that reduce inflammation without damaging stomach mucosa. However, a number of issues have arisen with these products, most notably with Vioxx, which recent studies have shown to increase the risk for heart attacks.[24]

Aspirin represents one of humankind's oldest pharmaceutical agents and continues to be a mainstay therapy for a variety of indications. Like all drugs, aspirin can be toxic at high doses (greater than 150 milligrams per kilogram body weight), but the benefits of aspirin clearly outweigh the risks. We might consider aspirin a true "wonder drug," as it has been shown to be useful in the treatment of a variety of conditions beyond fever and pain, including prevention of coronary artery disease, heart attack, and stroke. Recent studies suggest that aspirin may also limit the rate of growth and the occurrence of certain types of cancer, including prostate, colon, pancreatic, and lung cancer.[25] While new drugs will continue to treat these and other diseases, aspirin will always hold a significant place in the history of pharmaceutical agents.

Aspirin is believed to reduce the risk of a second heart attack by 20 percent.

1. A. A. J. Andermann, "Physicians, Fads and Pharmaceuticals: A History of Aspirin," McGill University School of Medicine, www.medicine.mcgill.ca/mjm/v02v02/aspirin.html (accessed September 25, 2007).

2. J. G. Mahdi et al., "The Historical Analysis of Aspirin Discovery, Its Relation to the Willow Tree and Antiproliferative and Anticancer Potential," *Cell Proliferation* 39 (2006), 147–155.

3. D. Jeffreys, *Aspirin: The Remarkable Story of a Wonder Drug* (New York: Bloomsbury, 2004).

4. Mahdi et al., "Historical Analysis of Aspirin Discovery" (cit. note 2), 148.

5. Ibid.

6. Aspirin Foundation of America, "Aspirin, 1999–2007," www.aspirin.org (accessed September 25, 2007).

7. Mahdi et al., "Historical Analysis of Aspirin Discovery" (cit. note 2), 148.

8. Ibid.

9. T. J. MacLagan, "The Treatment of Rheumatism by Salicin and Salicylic Acid," *Lancet* 1 (1876), 342.

10. W. Sneader, "The Discovery of Aspirin," *Pharmaceutical Journal* 259 (1997), 614–617.

11. Mahdi et al., "Historical Analysis of Aspirin Discovery" (cit. note 2), 149.

12. Andermann, "Physicians, Fads, and Pharmaceuticals" (cit. note 1).

13. Ibid.

14. Ibid.

15. Aspirin Foundation of America, "Aspirin" (cit. note 6).

16. P. J. Piper and J. R. Vane, "Release of Additional Factors in Anaphylaxis and Its Antagonism by Anti-Inflammatory Drugs," *Nature* 223 (1969), 29.

17. World of Molecules, "Aspirin," www.worldofmolecules.com/drugs/aspirin.htm (accessed September 25, 2007).

18. M. Hamberg, J. Svensson, and B. Samuelsson, "Thromboxanes: A New Group of Biologically Active Compounds Derived from Prostaglandin Endoperoxides," *Proceedings of the National Academy of Sciences U.S.A.* 72 (1975), 2,994.

19. J. R. McTavish, "The Industrial History of Analgesics: The Evolution of Analgesics and Anti-pyretics," in *Aspirin and Related Drugs,* ed. Kim D. Rainsford (London: Taylor & Francis, 2004), 25–44.

20. Piper and Vane, "Release of Additional Factors" (cit. note 15), 29.

21. Hamberg et al., "Thromboxanes" (cit. note 17), 2,997.

22. J. C. Collier and R. J. Flower, "Effect of Aspirin on Human Seminal Prostaglandins," *Lancet* 2 (1971), 853.

23. J. R. Vane and R. M. Botting, "The Mechanism of Action of Aspirin," *Thrombosis Research* 110 (2003), 255–258.

24. M. Elliott, M. D. Antman, and D. DeMets, "Cyclooxygenase Inhibition and Cardiovascular Risk," *Circulation* 112 (2005), 759.

25. Mahdi et al., "Historical Analysis of Aspirin Discovery" (cit. note 2), 147.

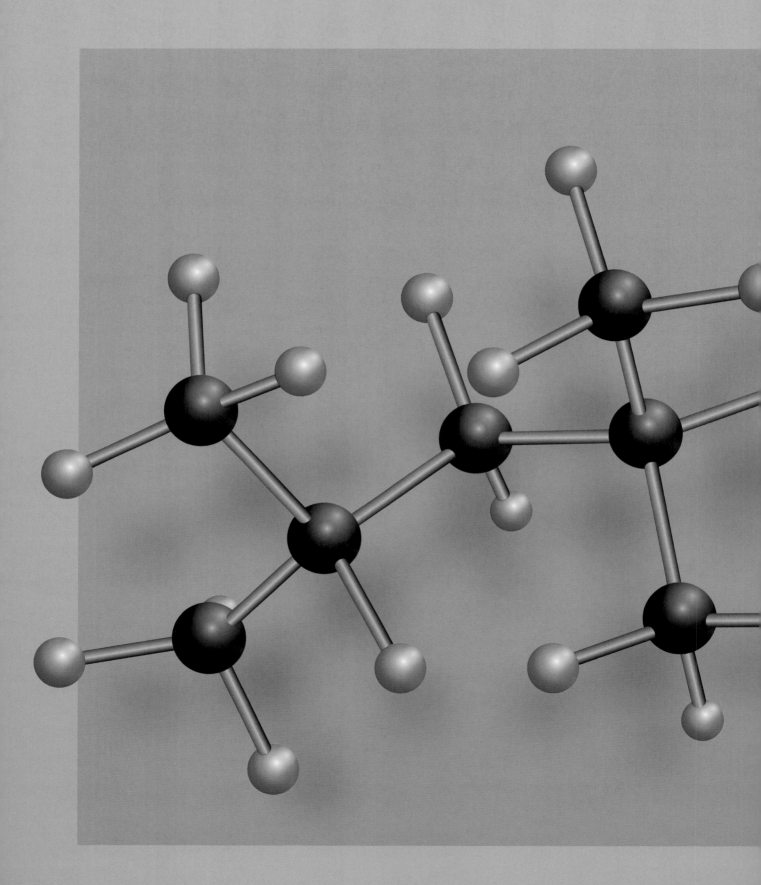

1910

ISOOCTANE

C_8H_{18}

U.S. PATENT NUMBER: 2,360,253
CAS REGISTRY NUMBER: 540-84-1

The octane ratings of gasoline visible at every gas station in the land—to which the compound isooctane is intimately related—were first defined in 1926 at the research laboratory of the Ethyl Corporation in Yonkers, New York. Why there? Two years earlier the company had been established to manufacture the recently synthesized gasoline additive tetraethyl lead. Following the relationship of these two historically linked compounds takes us back to the second decade of the twentieth century, when our society's love affair with the automobile and the gasoline-powered internal combustion engine was just beginning. This affair, for better or worse, continues to this day.

In the second NASCAR race of the 2007 season, megastar Dale Earnhardt, Jr., and his teammate dropped out early and finished at the back of the forty-three-car field at California Speedway. According to Earnhardt the team's engine builders had trouble with the new rule requiring that the high-octane fuel for NASCAR's premier series be lead (i.e., tetraethyl lead) free.

by Mary Ellen Bowden and Neil Gussman

For passenger cars driven on public roads, lead-free fuel became mandatory in the 1970s, but the NASCAR formula had preserved automotive technology in its 1950s form. Even today, NASCAR racing teams employ cast-iron block engines designed in the 1950s with carburetors—those complex, inefficient, fuel-air mixing devices that at one time sat atop every engine and now are found only in NASCAR stock cars, cars in Cuba, and smoky vehicles still being produced in a small number of former communist countries.

Why did lead, specifically tetraethyl lead, remain in NASCAR engines until 2007? The answer is performance. And the use of tetraethyl lead to achieve it was the brainchild of two innovative geniuses of the early twentieth century, Charles F. Kettering and Thomas Midgley, Jr.

At a time when electric street railroads and automobiles were being championed by Thomas Edison and others, Kettering's invention of the battery-powered starting engine, first installed in the 1912 Cadillac, helped secure the future for the gasoline engine. Many old movies capitalize on the humor of watching some poor fool attempt to start an automobile engine by hand cranking, but sometimes the re-

The length of a molecule of isooctane (about 1 nm) is to one inch as one inch is to the 16,000-mile Pan-American Highway, extending from Alaska to Panama.

ABOVE Ethyl Gasoline ad, *The Saturday Evening Post*, Ethyl Corporation, 1949
OPPOSITE Pegasus, Mobil gas station sign, c. 1935

sults were not funny. Kettering's boss broke his arm that way, and so Kettering put his mind to remedying the situation with a battery to power a small starting motor, but a battery far smaller and lighter than those needed to power a wholly electric car.

Another problem with the early automobile engine was "knocking," or preignition. At its worst, knocking can be violent enough to cause the self-destruction of an engine, and some early auto enthusiasts ascribed knocking to the new electric starting system. Kettering, however, suspected that the source was not the engine but the fuel.

In those days the mixture that made up gasoline was highly variable, and there was no standard from gas station to gas station. Gasoline then, as now, was made up of many compounds containing molecules with five to twelve carbon atoms apiece; the complex mixture varies according to the source of the petroleum and the refining processes used. Refining processes at the beginning of the twentieth century yielded less than 20 percent gasoline from a barrel of oil. The rest became kerosene and lubricants, and much of it was just discarded.

Kettering turned to finding a gasoline additive that might reduce knocking. When he moved from his Dayton Engineering Laboratory (Delco) to General Motors in 1916, he handed the research over to a recent hire at Delco (soon to become a subsidiary of GM), Thomas Midgley, a bright young engineer from Cornell University. Midgley invented the "bouncing-pin" test engine so that knocking could be easily determined by ear. He also put together an optical gas-engine indicator to magnify and record the shape of the pressure wave resulting from the combustion of fuel and air inside the engine's cylinder.

When the United States entered World War I in 1917, creating an antiknocking fuel for airplanes, which still used internal combustion engines, became a national priority. When an automobile developed

Before 1950 coal was the principal fuel source in the world. After 1950 petroleum became the top fuel source.

serious knocking, the driver could simply steer it over to the side of the road. An airplane pilot sensing knocking did not have such an option and could soon be in danger. But the solution to knocking was slow in coming.

After trying many potential additives, Midgley and his group began systematically investigating organometallic compounds—compounds whose molecules are made up of atoms of carbon and hydrogen, like the molecules in gasoline, plus metal atoms. Tetraethyl lead appeared to be the ideal solution, at least in 1922.

Organic compounds of lead do not burn efficiently. A car burning leaded gas actually delivers slightly less power per gallon of gas than one burning unleaded gas. But what made tetraethyl lead vital to internal-combustion-engine designers in the 1920s was its ability to prevent preignition of the fuel inside an engine. In a four-stroke cycle engine that uses gasoline for fuel, a mixture of fuel and air is drawn into the engine as each piston travels down in the cylinder and then is compressed as the piston travels up. At a predetermined point, just before the piston reaches the top of its upward travel, a spark plug in the cylinder ignites the fuel-air mixture, causing the release of expanding gases. These propel the piston back down and deliver power to the crankshaft and on to the wheels.

One method of making more power for a given size of engine was to increase the compression ratio—the ratio of the capacity of the cylinder to the volume of the fuel-air mixture when it is compressed. Early engines used compression ratios in the range of six to one. Raising this ratio increased power but had a huge drawback. When the fuel-air mixture is compressed to less than a tenth of its original volume in a hot cylinder, a portion may preignite, interfering with the power

Diagram of tetraethyl lead

Robert Dawson
Homes Refinery, Long Beach, CA, 1993
Gelatin silver print
16 x 20 inches

Robert Dawson
Aerial View of Oildale, CA, 1991
Gelatin silver print
16 x 20 inches

surge caused by the rest of the fuel igniting. The result of this interference is knocking and lost power.

Another part of the solution to the problem of knocking was to rate gasolines on their characteristics to produce knocking—or more to the point, not to produce knocking. This knowledge would be important for any automobile engine designer or car owner. At Ethyl Corporation they had to figure out how much tetraethyl lead to add to various gasolines to inhibit knocking.

When Russell E. Marker, a maverick organic chemist still in his twenties, arrived at Ethyl's research laboratory in 1926, it was just a renovated garage. Graham Edgar, the director of the Ethyl Corporation's lab, set Marker to work on the second floor synthesizing more organometallic compounds. All day long Marker heard the annoying sound of the knocking of the Delco bouncing-pin test engines coming up from the first floor. There a team of researchers was developing a test to rate gasolines. Marker ambled downstairs to chat.[1]

According to Marker, he suggested the team use hydrocarbon compounds with nearly the same boiling points in the test gasoline. The team had been encountering problems with its test gasolines because the knocking characteristics changed when different portions of the mixture vaporized at different times. So Marker first synthesized pure normal heptane (with seven carbons to the molecule), which caused severe knocking, almost cracking the cylinders of the test engine. Then he synthesized an octane called isooctane, with eight carbons arranged in branched formation, which caused no knocking at all. Marker concluded erroneously that the difference in characteristics of the two hydrocarbons had simply to do with even versus odd numbers of carbon atoms in the molecules (eight versus seven). When he tried normal octane, however, that too

Isooctane is a reference substance for determining octane number or performance standard in gasoline. The octane number is the number you see on the pump.

Diagram of n-heptane

Isooctane
Installation view, Tang Museum

caused knocking. Subsequent fuel scientists discovered that the branched structure of isooctane molecules acts to inhibit preignition.

The team then created several test gasolines made up of different percentages of pure isooctane and pure heptane. These mixtures closely replicated the amount of knocking produced by any gasoline being rated. When we see the number 87 on a gas pump for "regular" gasoline, with its variable additives, it performs the same as a standard gasoline made up of 87 percent isooctane and 13 percent heptane. And "high-test" means that particular gasoline closely compares to a test gasoline with approximately 91 or 93 percent isooctane.

That said, the scientific truth is even more complex, because American octane ratings stated at the pump are an average of octane ratings measured in test engines under two controlled but different sets of conditions—with the engines under less and more stress.

Adding tetraethyl lead to gasoline allowed refineries of the 1930s to create high-octane fuel for high-compression engines, fuel that they could not produce using the refinery technology of the time. The primary benefit of high compression comes at the peak of an engine's power. In most cases cars use only a small fraction of their available power and do not need high-compression engines. So who wanted high-compression, high-performance engines? People who ran their engines at or near full power very often—race-car drivers and military pilots to name just two. In the 1930s and 1940s every major combatant in World War II was anticipating the conflict to come and designing high-performance aircraft. The big high-performance engines had to be both powerful and reliable, which meant balancing high compression with the potential damage from knocking. Aviation gas was highly leaded, with octane ratings so great the fuel was not sold for passenger-car use. In fact, car racers bought "av

ABOVE Diagram of isooctane
OPPOSITE Mobil Oil and Gasoline advertisement
for antiknocking gasoline, c. 1950

gas" for their cars so they could increase the compression ratios of their engines.

Many of the legendary fighter aircraft of World War II would have been much less effective without leaded gas. And automotive marketers after the war found that horsepower was sexy. Horsepower, the calculation of the maximum power that can be delivered by an engine at a given engine speed, became the number that a marketer could use in a simple, "more is better" fashion to link their shiny sedan to the great speedways at Indianapolis, Monza, and Le Mans. Higher compression was the route to higher horsepower, so the top-of-the-line engines began to be made with higher compression ratios requiring higher-octane premium fuel.

Collectively, Americans drive seven billion miles a day.

Is horsepower a good measure of performance? The answer is yes, if you are talking about maximum performance. In a fighter aircraft climbing and twisting to escape an enemy, more horsepower can be a matter of life and death. For a racer holding his foot flat on the floor on the long straights at Indy and Le Mans, more horsepower can mean more wins. But horsepower is not a very useful measure of how a car will perform in the varied circumstances of driving on roads and highways.

By the 1960s, the heyday of high-performance engines, the big three carmakers of that time (GM, Ford, and Chrysler) and several smaller ones each had an engine tuned for maximum horsepower

Ed Ruscha
Mocha Standard, 1969
Color screen print on mold-made paper, edition of 100
24⅞ x 40 inches

that was winning races at drag strips and speedways. But the street versions of these speedway monsters were expensive, required a lot of maintenance, and were difficult to drive. They had power—more than 400 horsepower was the minimum in this league—but they got gas mileage that was noticeably poor even in an era of gas guzzlers, and they also required a skilled pair of feet on the gas pedal and the clutch.

Automobile marketers were in fact using a bait-and-switch trick to lure buyers into dealerships. The 425-horsepower L-88 Corvette enticed boys of all ages into the Chevrolet showroom, drooling for the Corvette's big, inefficient, high-performance engine, but most of them left with a mundane Malibu. And Plymouth sold hundreds of 335-horsepower base-model Road Runners with the "beep-beep" horn for every 425-horsepower Hemi Road Runner Superbird.

By 1973 every new car sold in America could run on regular gas (87 octane), and tetraethyl lead was on its way out of gasoline formulations—except for those used in propeller-driven aircraft and NASCAR stock cars. Studies had shown that lead—often small amounts absorbed over a period of time—does in fact have serious health effects, especially on the brain. Meanwhile, through various new technologies developed in the 1930s and afterward, petroleum refineries were able to create gasolines with antiknocking characteristics without the need for tetraethyl lead. By catalytic processes they broke down long molecules and reformed others into high-octane compounds. Today petroleum refiners can get far more gasoline out of a barrel of crude oil than they could at the beginning of the gasoline-powered automobile era— now about 50 percent. The number of other petroleum products derived from crude oil has increased fantastically; almost nothing is thrown away.

Plymouth Road Runner Superbird, 1970

$$C_8H_{18} + \frac{25}{2}O_2 \rightarrow 8CO_2 + 9H_2O$$

Isooctane
Installation view, Tang Museum

In 1996 the United
States consumed
approximately
45 percent of the
petroleum produced
worldwide that
year.

From the 1950s onward the air pollution caused by millions of cars and trucks was a recognized problem. Eugene Houdry, the inventor behind Sunoco's catalytic cracking units that supplied high-octane gas for military aircraft in World War II, designed catalytic converters for automobiles to change incompletely combusted hydrocarbons, carbon monoxide, and nitrogen compounds spewing from automobile exhaust into more benign end products: nitrogen, water, and carbon dioxide. Because such converters would clog with lead, emissions standards and the installation of catalytic converters on new cars had to await lead's removal. What was not at first realized is that carbon dioxide and water vapor are not without consequences: they are the chief cause of global warming.

Since the 1990s oxygenates have been added to gasoline to further the complete combustion of the fuel by contributing more oxygen atoms to the gases igniting in the engine and thereby reducing air pollution from cars. These compounds also raise octane ratings. Since they are already partially oxygenated (in a sense combusted), they do not contribute to preignition. For a similar reason they never raise gas mileage. The oxygenate compound MTBE (methyl tert-butyl ether) was favored for over a decade but has a significant drawback: it is carcinogenic, although probably not to a great extent. However, since MTBE is very soluble in water, the chances of it leaching from gasoline spills or leaks into groundwater are reasonably high.

Ethanol, made from the fermentation of corn, has recently been hailed as the new answer to pollution, although the jury is still out on whether more or less energy is expended in making this fuel than in refining petroleum. Further, which product costs more in dollars and cents depends on what is counted among the costs. Incontrovertibly,

Diagram of MTBE (methyl tert-butyl ether)

ethanol comes from corn, a renewable resource grown domestically—a clear attraction, especially when the regions of the world that have the largest known oil resources are in upheaval.

Regardless of whether ethanol or some other fuel eventually displaces gasoline as we know it, isooctane has well and truly parted company with tetraethyl lead—even in NASCAR vehicles. And we have almost come full circle in our notion of the most desirable automobile engine. Some would say it is an electric motor.

1. Marker, now deceased, has given his own account of the derivation of octane ratings in an oral history held at the Chemical Heritage Foundation in Philadelphia that differs in some details from Edgar's story.

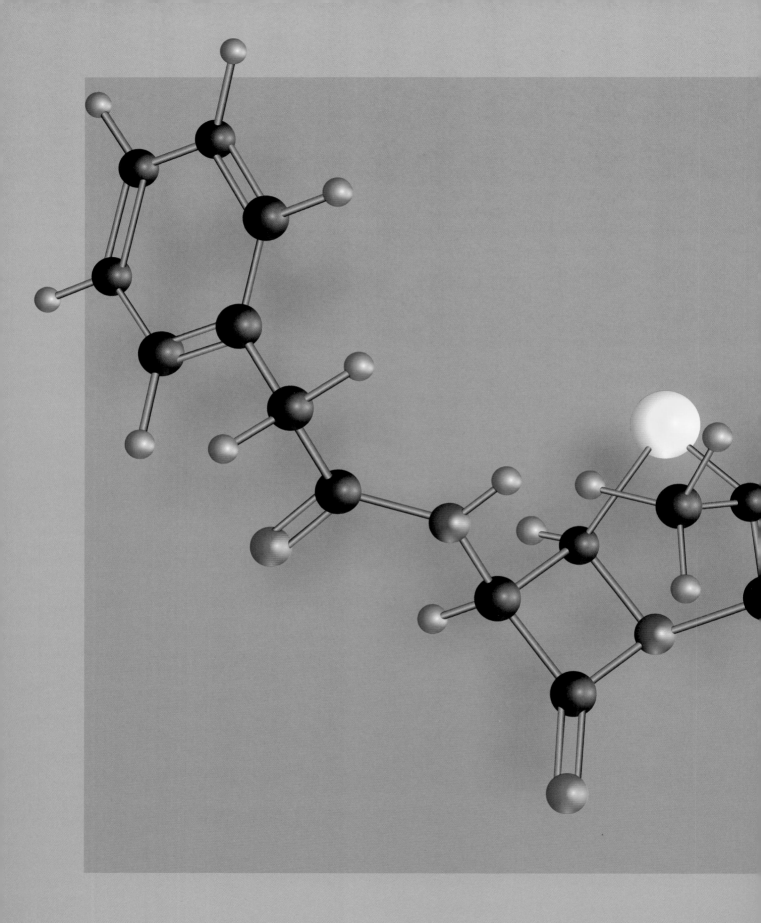

PENICILLIN G

$C_{16}H_{18}N_2O_4S$

U.S. PATENT NUMBER: 2,480,466
CAS REGISTRY NUMBER: 113-98-4

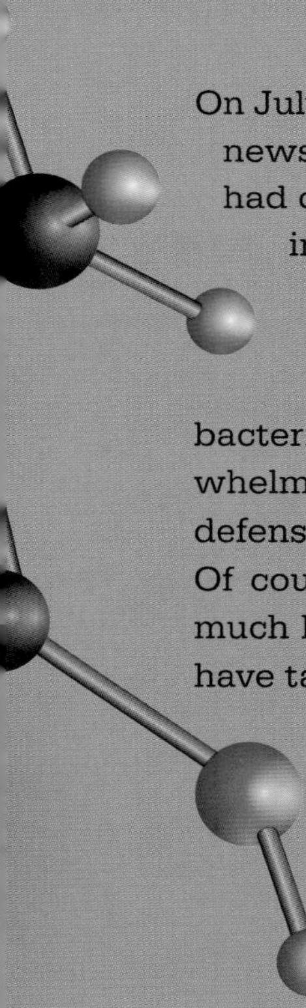

On July 7, 1924, President Calvin Coolidge received the heartbreaking news of the death of his sixteen-year-old son. Coolidge's namesake had died as a result of an infected foot blister sustained while playing tennis with his brother just eight days earlier.[1] Coolidge's sadness must have been exacerbated by the fact that, despite his power as president, he could not save his son from this tragic fate. But such was the nature of life at that time: common bacterial infections were potentially lethal, capable of simply overwhelming individuals whose natural immune systems were their only defense. Physicians had no effective means to counter such infections. Of course, penicillin would not be discovered until 1928, and only much later was it developed into the array of modern antibiotics we have taken for granted since the 1960s.

Most Americans first learned of penicillin as a result of another human tragedy—the Cocoanut Grove fire that occurred in Boston on the night of November 28, 1942, claiming almost 500 lives and injuring hundreds more.[2] The Cocoanut

by Raymond J. Giguere

Grove was a popular nightclub packed that evening with over a thousand people when a fire spread rapidly throughout the building. The U.S. government seized on this civilian disaster to mobilize its emergency response teams in an effort to save lives and to test readiness for a potential foreign military attack. At that time penicillin remained a closely guarded military secret, and studies of the drug were still in the preclinical phase.[3] Despite the drug's scarcity and cost, top-level administrative orders were given calling for the release of all the penicillin that American companies could spare. John Sheehan, then a young organic chemist at Merck & Company under the direction of the renowned researcher Max Tischler,[4] described how he and his coworkers at Merck labored "around the clock, in relays" to isolate and purify all Merck's crude penicillin. Police then escorted the priceless payload to Massachusetts General Hospital to aid the fire victims.[5] Sheehan estimated that many more lives would have been lost as a result of burn infections had penicillin not been pressed into service. Even so, penicillin remained in such short supply that Tischler reminded his coworkers why only minuscule amounts were released for experimentation: "Remember, when you are working with 50 or 100 milligrams of penicillin, you are working with a human life."[6]

Although the government tried to keep this emergency penicillin response from the public, the news media succeeded in uncovering the story. And imploring letters to President Franklin D. Roosevelt, such as this one from late 1943, soon poured in by the hundreds:

> Dear Mr. President:
> I know you are busy with the war, but could you spare a few minutes and read my letter? I am in great need of your help. My husband is in great need of the new drug penicillin. An infection has settled in his blood stream and I just got a call from the

Penicillin was first tested on mice with great success and was immediately tested on humans. Penicillin is toxic to guinea pigs, another common test animal.

Frank Moore
Beacon, 2001
Oil on canvas over featherboard
72 x 96 inches

doctor saying my husband has turned for the worst. I have two sons serving Uncle Sam, one is a sergeant in the Army Air Corps and my other son is in the Navy and is somewhere at sea. His ship sunk a German sub about a month ago.

Mr. President, I gave Uncle Sam two sons, could you please try and get me that new drug and save my husband? If you can get me the drug, have it sent to Church Home Hospital right across from Johns Hopkins Hospital on the Broadway Street entrance. His name is Louis G. Profili, fourth floor, room three. I'm almost out of my mind. If you please, try and help me before it's too late. I don't know how to thank you, because words can't express what is in my heart.

Thank you,

[Signature marked out—presumably Mrs. Profili][7]

It is touching that the American people had such faith in the power and compassion of their leader, as well as in the emerging reputation of penicillin as a "magic bullet." Unfortunately, very few of these many personal requests were ever answered or fulfilled. In the early 1940s the public had extremely limited access to this "wonder drug." Until mid-1944 penicillin was so new and precious that its use was restricted to the military and to ongoing scientific research.[8]

Some scientists estimate an average of ten years has been added to humans' life span in the twentieth century as a result of penicillin development.

The history of penicillin—from Alexander Fleming's contaminated petri dish to its status as one of the most sought-after remedies in history—is a long, complex, and fascinating story, aspects of which we can only touch on here. The path of its development led to top-secret military alliances, to intense research efforts on both sides of the Atlantic to unlock its chemical makeup, and shortly after World War II to a period of resignation in the research community regarding the possibility of ever synthesizing penicillin.

In 1928 Fleming, a bacteriologist working at St. Mary's Hospital in London, discovered penicillin through a lucky accident. Fleming was growing staphylococci in petri dishes, and "he noticed that the mold had spoiled one of his cultures. Staphylococcus grew on only half of the plate. . . . He noticed that the mold had cleared a wide, bacteria-free area between itself and the staphylococci—perhaps had killed them."[8] What happened next was critical to the fate of millions

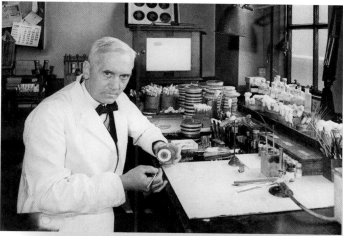

whose lives were saved by Fleming's observation. Fleming salvaged the petri dish, wondering whether the mold was producing a substance responsible for killing the deadly strain of bacteria he was intentionally growing. His hunch turned out to be correct; the mold was the source for research into what became the penicillin family of antibiotics. Interestingly, strong evidence exists that the observation Fleming made was not a unique or even an unusual event. Molds had been used previously for combating infections.[9] But Fleming's training and experience (including discovery of another antibacterial, lysozyme, present in human saliva), allowed him to look at a "failed" culture dish and see potential for a drug that proved so fruitful for humanity.[10] As Louis Pasteur, the great nineteenth-century bacteriologist, once mused: "Chance favors the prepared mind."

Fleming shared the 1945 Nobel Prize in Physiology or Medicine with Ernst Chain and Howard Florey, two scientists who researched penicillin in Great Britain. A decade passed between Fleming's 1929 publication on the discovery and Florey's research group at Oxford

Alexander Fleming with a petri dish of staphyloccoci and penicillin, c. 1920

Jean Shin
Chemical Balance 2, 2005
Prescription pill bottles, mirrors, and epoxy
7 units from 32 to 46 inches in
diameter, overall dimensions variable

Penicillin
Installation view, Tang Museum

University beginning its independent study of penicillin as a potential agent for combating bacterial infections. Some scientists felt that Fleming had been awarded too much credit for his role in this amazing discovery and its early development. Fleming did not conduct any early human tests using penicillin; his lab and the financial and political support he had at that time did not provide him the means to do so.[11] Moreover, early penicillin research was extremely difficult; isolating and purifying sufficient quantities from the mold, and storing and handling penicillin in a manner that would not cause it to lose its medicinal properties were daunting challenges. Ernst Chain, a brilliant German chemist at Oxford, made significant contributions to purification and isolation techniques, such as using freeze-drying (then a new technique) to improve isolation of the molecule from the mold.

Approximately 10 percent of the population is allergic to penicillin.

Early chemists who studied penicillin were not even certain of its precise chemical structure, making manufacturing penicillin even more difficult. Without knowledge of the correct molecular structure, penicillin researchers were unable to synthesize the new drug. Chemical synthesis offered an alternative to culturing the mold and the painstaking problems of then isolating the molecule. Fortunately, in 1943 breakthroughs in applying deep-fermentation techniques (developed largely at the Northern Regional Research Laboratory, now the National Center for Agricultural Utilization Research, in Peoria, Illinois) allowed researchers to grow the mold rapidly, providing the short-term answer to the penicillin production problem.

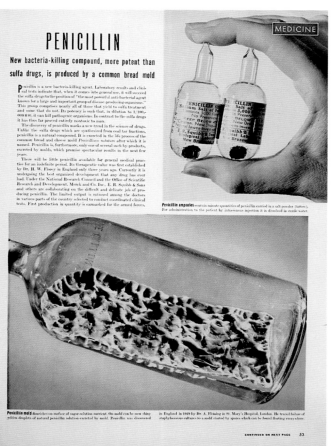

In May 1943 *Life* magazine ran its first story on penicillin, complete with photos of surface-fermentation jars and commercial labs. There was no mention of the deep-fermentation techniques under development; perhaps the *Life* article was a decoy to distract the Axis powers by misleading them with an inefficient means of penicillin mass production.[12] Surprisingly, Schenley Laboratories publicly proclaimed just one year later in a prescient magazine advertisement for the new drug: "When the thunderous battles of this war have subsided to pages of silent print in a history book, the greatest news event of World War II may well be the discovery and development—*not* of some vicious secret weapon that *destroys*—but of a weapon that *saves* lives. That weapon, of course, is penicillin." That same year *Time* magazine's cover featured Alexander Fleming, discoverer of penicillin, and hailed the breakthrough of deep-fermentation methods. The new approach boosted production in the United States to nearly nine pounds per day, or "enough to treat 250,000 serious cases a month."[13] In contrast, penicillin production in all of 1943 was estimated at just fifteen pounds.[14] Fueled by the new production method, the role penicillin played in reviving infected soldiers is arguably one of the principal reasons the Allies prevailed in World War II.

In 1945 Dorothy Hodgkin, an Oxford University researcher, determined the chemical structure of penicillin, a critical accomplishment that had eluded scientists for well over a decade. This achievement helped

The secret Allied World War II effort to synthesize penicillin involved thirty-nine labs and over a thousand chemists at its peak; the project failed, and the synthesis of penicillin became known as "the impossible problem."

OPPOSITE "Penicillin," May 24, 1943, excerpt from *Life* magazine
ABOVE *Time* magazine cover of Alexander Fleming, May 15, 1944

earn her the 1964 Nobel Prize in Chemistry and provided the molecular foundation for continued efforts aimed at synthesizing penicillin, which had been produced since 1944 through deep fermentation.[15] In 1957, John Sheehan and his research group at the Massachusetts Institute of Technology accomplished the molecule's chemical synthesis. Sheehan's success led to a host of new penicillins and a rebirth of interest in the field.

Penicillin was only the beginning of the antibiotic revolution. Consider that in 1900 five of the ten leading causes of death in the United States were from infectious diseases that could be treated effectively with penicillin and related antibiotics. At the top were pneumonia and influenza; tuberculosis; and diarrhea, enteritis, and ulceration of the intestines. By 2000 only one of these—pneumonia and influenza—remained on the list, and it had dropped to seventh position. Of course the reduction in infectious diseases also resulted from better purification methods for drinking water that were in place by the 1920s. Nonetheless, the number of lives saved through the development of the penicillin antibiotic family is immeasurable: some historians estimate that up to one-third of any group of people living in modern nations today would have died

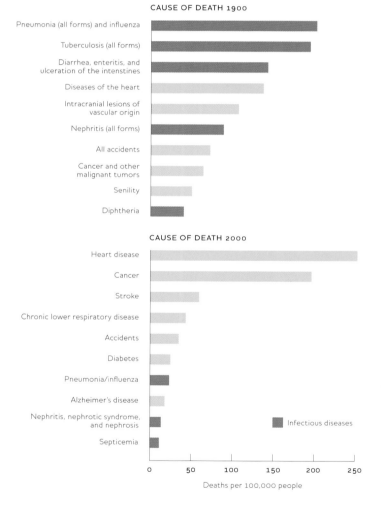

CAUSE OF DEATH 1900

Pneumonia (all forms) and influenza
Tuberculosis (all forms)
Diarrhea, enteritis, and ulceration of the intenstines
Diseases of the heart
Intracranial lesions of vascular origin
Nephritis (all forms)
All accidents
Cancer and other malignant tumors
Senility
Diphtheria

CAUSE OF DEATH 2000

Heart disease
Cancer
Stroke
Chronic lower respiratory disease
Accidents
Diabetes
Pneumonia/influenza
Alzheimer's disease
Nephritis, nephrotic syndrome, and nephrosis
Septicemia

■ Infectious diseases

0   50   100   150   200   250

Deaths per 100,000 people

•

Leading causes of death in the United States in 1900 and 2000

from childhood infections without antibiotics.

The 1960s witnessed the growth in use of semisynthetic penicillins and related antibiotics; their misuse and overuse realized a key challenge that Fleming predicted in the 1940s—the rise of antibiotic-resistant bacteria.[16] The development of penicillin G–resistant bacteria, which was noted within a few years of the drug's introduction, results from the process of natural selection. Bacteria reproduce very rapidly (some in as little as thirty minutes) and under evolutionary pressure develop mutations that make them resistant to antibiotics. Hospital environments offer nearly optimal conditions for such mutation, owing to the frequent contact between bacteria and antibiotics. Consequently, these challenges require ongoing research in medicinal chemistry to discover and develop new molecular agents, as well as education in the public health arena to alert us to the dangers of overuse and overprescription. When considering the tension surrounding this timely issue, the last lines of Robert Bud's definitive book *Penicillin: Triumph and Tragedy* come to mind: "Will it end with the final dramatic phase: *catastrophe*? At the

Penicillin advertisement, Schenley Laboratories, Inc., c. 1944

Penicillin was taken for years by patients before its chemical structure was identified.

beginning of the twenty-first century, the prognosis is still uncertain."[17] That said, this precarious situation reminds us of the need for continued vigilance when dealing with human disease and its causes, as unintended consequences accompany human progress. In 1900 the world population was around 1.6 billion; by 1950 it was 2.5 billion; and in 2000 it was over 6 billion. Many lives have been saved by penicillin and other antibiotics, but the lingering question of sustainability increasingly needs to be addressed as we assess whether we will one day become victims of our own success.

1. Dr. Zebra, "The Health and Medical History of President Calvin Coolidge," www.doctorzebra.com /prez/t30.htm (accessed July 6, 2007).

2. John C. Sheehan, *The Enchanted Ring: The Untold Story of Penicillin* (Cambridge, MA: MIT Press, 1982), 40–43.

3. Ibid., 44–57.

4. Sheehan would become a professor of organic chemistry at MIT, and in the 1950s he led a graduate research group in completion of the first rational chemical synthesis of penicillin.

5. "Drug Rushed Here to Aid Fire Victims," *Boston Globe*, Dec. 2, 1942, 15.

6. Sheehan, *Enchanted Ring* (cit. note 3), 42. One hundred milligrams is only one-tenth of a gram. There are 28 grams in an ounce.

7. University of Pennsylvania School of Arts and Sciences, "Health, Medicine, and American Culture, 1930–1960–Penicillin: Public: Letters," ccat.sas .upenn.edu/goldenage /state/pub/letters/pages /letter_re_Profili.htm (accessed July 6, 2007).

8. "Penicillin," *Life*, July 17, 1944, 57-62.

9. Eric Lax, *The Mold in Dr. Florey's Coat* (New York: Henry Holt, 2005), 24–26.

10. "20th Century Seer," *Time*, May 15, 1944, 62.

11. Sheehan, *Enchanted Ring* (cit. note 2), 26–27.

12. "Penicillin, New Bacteria-Killing Compound, Is Produced by Bread Mold," *Life*, May 24, 1943, 53–55.

13. "Penicillin" (cit. note 8), 57.

14. "20th Century Seer," (cit. note 10), 61.

15. Nobel Prize Internet Archive, "Dr. Dorothy Crowfoot Hodgkin: Chemist, Crystallographer, Humanitarian (1910– 1994)," www.nobelprizes .com/nobel/chemistry /dch.html (accessed July 6, 2007).

16. Christopher Walsh, *Antibiotics: Actions, Origins, Resistance* (Washington, DC: ASM Press, 2003), 91–95.

17. Robert Bud, *Penicillin: Triumph and Tragedy* (New York: Oxford University Press, 2007), 216.

Penicillin and buckyball
Installation view, Tang Museum

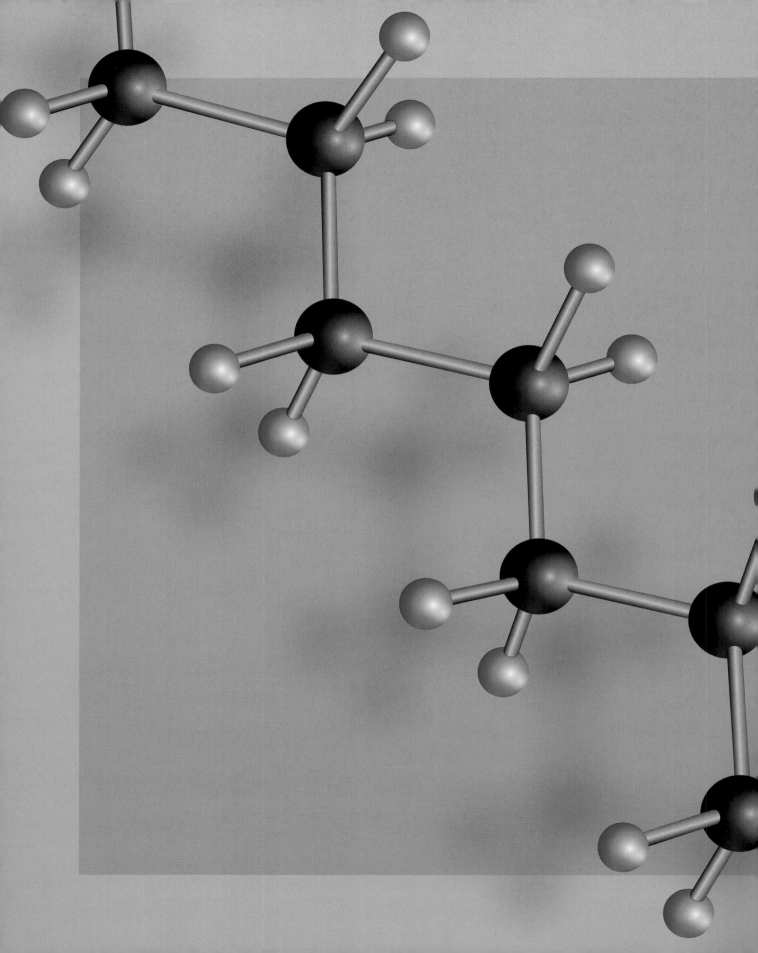

1930

POLYETHYLENE

$$(CH_2CH_2)_n$$

U.S. PATENT NUMBER: 2,232,475
CAS REGISTRY NUMBER: 9002-88-4

For centuries the natural limitations of wood, stone, and metals made it impossible to conceive of material desires beyond the traditional. That situation changed in the twentieth century when chemists learned to synthesize substances that had never before existed and to specify their properties. Because we are now surrounded by synthetic materials, it is difficult to recall how radically such plastics as polyethylene extended the limits of the possible. We live and work with machines, appliances, and furnishings whose visual appearances and tactile qualities our ancestors would have considered unnatural, even alien. Objects molded, extruded, or foamed from plastics have proliferated so much since the middle of the twentieth century that we now take them for granted. Synthetic materials have been naturalized.[1]

Early imitative applications of plastics exhibited pride in the ingenuity of illusion. Celluloid, introduced around 1870, imitated the layering of ivory, the mottling of tortoiseshell, and the veining of marble. But the qualities of plastics extended beyond surface imitation. Promotion of various types early in the twentieth century emphasized the

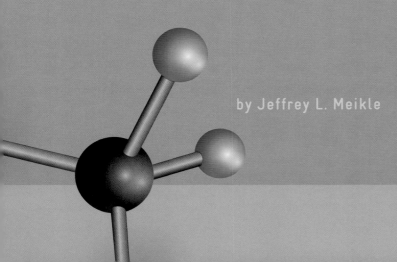

by Jeffrey L. Meikle

substitution of inexpensive chemical substances for scarce natural materials. Modern science was making former luxury goods available to democratic man and woman. Newer synthetic plastics like Bakelite, commercialized during the 1920s, often seemed superior to natural materials with their irregularities and flaws. Only superficially imitative, new plastics seemed miraculously artificial. The business magazine *Fortune* celebrated plastic in 1936 as "a new substance under the sun," which improved on, or even transcended, nature.[2] Over the decades this artificiality has changed our perception of reality by making the material world seem more malleable, less permanent, even ephemeral. Plastics are easier to shape and to color than any other material and have given everyday life a sense of greater material possibility. To be able to specify the stuff of existence at the molecular level; to imbue it with properties, textures, and colors unknown to earlier generations; to mold from it objects and environments unknown to prior civilizations—all these marked an unprecedented degree of human control over the material environment.[3] A retired DuPont chemist predicted in 1988 that humanity would eventually "perish by being smothered in plastic."[4] However, plastics held out a promise of material freedom in a wholly malleable environment. This duality runs throughout the American experience of plastics, sometimes as undercurrent, sometimes as conscious statement.

The post–World War II baby boomers grew up as the plastics industry was coming into its own. Before the war the use of plastics in consumer goods remained limited to a few products, such as celluloid dresser sets and Bakelite radios. Although the new industry's publicists predicted a utopia molded out of "miracle materials" derived from coal, water, and air, synthetic materials did not attain wide application until after the war. Annual production of plastics exceeded six billion pounds by 1960.[5] Dominance in the plastics

The word *plastic* comes from the Greek word *plastikos,* which means "to form."

Roxy Paine
S2-P2-BK1, 2006, *S2-P2-BK2*, 2006,
S2-P2-BK10, 2006, *S2-P2-BK15*, 2006,
S2-P2-BK18, 2006, *S2-P2-BK23*, 2006
Low-density polyethylene
28 x 22 x 24 inches

industry shifted away from thermosetting varieties like Bakelite, which was the hard, nearly indestructible product of an irreversible chemical reaction, to such thermoplastics as polyethylene, which was softer and less dense and could be melted and remolded indefinitely. The result was a flood of new products—garbage pails, squeeze bottles, hula hoops—that were lighter, more flexible, and less permanent than objects made from thermosets. The chemical feedstocks of the industry also changed. After a serious shortage of benzene, which was obtained as a by-product from coking coal, material suppliers turned to petroleum and natural gas for raw materials. The shift from coal to petroleum—from a dense solid to a viscous liquid—symbolized the shift from permanent thermosets to endlessly changeable, shape-shifting thermoplastics. Affluent postwar Americans, characterized by historian David M. Potter as the "people of plenty," had no trouble consuming an ever-increasing quantity of plastics, materially and visually contributing to a culture that was shapeless, ever-changing, and increasingly impermanent.[6]

More than sixty thousand different types of plastics are currently in production worldwide; since the 1950s, more polyethylene has been produced than any other plastic.

Industry publicists presented a vision of the world that was almost impossible to imagine. In 1957 *Life* magazine posited a "new world" in which "man makes a multitude of new things" using "materials whose nature he could not even have guessed at a few decades ago." Expressing a certain uneasiness, *Life* admitted that "even the familiar items have undergone such outlandish transformations" that one "can hardly recognize them."[7] The British design critic John Gloag lamented that plastics enabled overly imaginative designers to "lose self-control." Fascination with plastics was "not so much a state of mind as a state of intoxication" leading to "widespread misuse." According to Gloag, designers working with new plastics had embraced "an orgy of ornament" and were busily involved in creating "a new rococo period" marked by extravagance and vulgarity.[8] The postwar explosion of amoeboid, boomerang, and saucer shapes, all easily molded of plastics, contributed to a vigorous popular culture marked by an exuberant surrealism. As a historian of design later observed of postwar consumer products, "Commonplace objects took extraordinary form, and the novel and exotic quickly turned commonplace."[9] James Sullivan, a founder of Union Products, a molding company in Leominster, Massachusetts, recalled that in 1946 "virtually nothing was made of plastic and *anything* could be"—including the pink flamingos for which his company eventually became well known.[10] That versatility explained why plastics rapidly became indispensable.

It is estimated that only 3 to 5 percent of used plastic is recycled.

No plastic contributed to this trend more than polyethylene, a soft, waxy, flexible thermoplastic that the British firm Imperial Chemical Industries (ICI) had developed during the late 1930s and had applied to electrical insulation and radar shields for aircraft during the war. "It was a polymer so unlike the polymers known at the time," a researcher at ICI recalled, that at first "no one could envisage a use for it."[11] By 1952, however, when ICI and DuPont lost monopoly control of polyethylene in an antitrust judgment, it seemed the most promising new plastic. Construction of plants by eight other American chemical companies (including Allied Chemical, Dow, Eastman Kodak, Monsanto, and Phillips Petroleum) created a buyer's market for a plastic from which an optimistic molder claimed it was "difficult to mold a bad product."[12] The price of polyethylene molding

OPPOSITE Raw polyethylene being blow molded to form plastic bags, Eastern Packing, Leominster, MA
RIGHT Don Featherstone pink flamingos, Union Products, Inc., Leominster, MA, 2006

resin fell quickly from an initial forty cents a pound to less than a dime—enabling entry into a host of new markets, whose demands absorbed production and in turn stimulated speculative expansion. By 1958 polyethylene producers were churning out the industry's highest annual volume to date, 920 million pounds, and had a growth rate of 30 percent.[13]

In becoming the first of the mass plastics, polyethylene changed the way people thought about synthetic materials. Although Bakelite had gained acceptance in the 1920s and 1930s through its durability, polyethylene soon became an archetypal plastic of a different sort, as oversupply led producers to think not of durability but of disposability. Much of the never-ending flow of polyethylene wound up as packaging. Soft and flexible, the material yielded the innovative squeeze bottle, first used in 1947 for Stopette deodorant. Three years later the Plax Corporation was blow-molding a hundred thousand squeeze bottles a day and planning to triple capacity to mold unbreakable containers for hand cream, suntan lotion, shaving cream, shampoo, and cologne. Polyethylene was so successful that a consultant urged Procter & Gamble to use it as packaging for Crisco shortening. A blow-molded polyethylene bottle was initially several times as expensive as glass but fell rapidly in price, and inexpensive blow-molded thermoplastic bottles were soon used for detergent, bleach, milk, sauces, and condiments.

Twenty-three countries have banned, taxed, or restricted the use of plastic bags because they clog sewers and lodge in the throats of livestock.

Polyethylene appeared in other highly visible, equally disposable consumer goods such as cheap toys. Young boys played war with realistic toy soldiers injection molded from dark-green polyethylene or recreated the days of Fort Apache with cowboys and Indians molded with so little authenticity that a chief in full regalia might appear in solid purple. Young girls played dress-up with stringless loops of pop beads, each molded with peg and socket to pop into flexible connection with another bead on either side. Offered initially with a

Dan Peterman
Tische, 1996
Recycled, post-consumer plastic
15 × 44 × 19 inches

Tony Cragg
New Figuration, 1985
Plastic wall construction
113 x 54 inches

pearly lacquer as novelty jewelry for adults, the beads soon appeared in gaudy popsicle colors aimed at appealing to children. Although pop beads consumed forty thousand pounds of resin each month in 1956, their economic and cultural resonance was eclipsed by the hula hoop, one of the first fads of baby boomers. Sold for about two dollars by Wham-O, a California molding company that later marketed Frisbees, hula hoops were so popular they inspired "chiselers" and "jackals" to offer badly assembled knockoffs for fifty cents. The twirling hoops absorbed about fifteen million pounds of a new high-density polyethylene resin, one third of the total output for 1958, that otherwise would have served witness to the folly of overproduction.[14]

One of the earliest toys made out of polyethylene was the hula hoop, invented in 1958. One-third, or about fifteen million pounds, of the high-density polyethylene produced that year was used to construct hula hoops.

This new high-density, or linear, polyethylene was a marked improvement on the basic type developed before the war. Polymerization of ethylene had remained difficult owing to the extreme pressures required, up to 30,000 pounds per square inch. A search for catalysts of low-pressure polymerization ended with several near simultaneous breakthroughs in 1953 and 1954. Desire for an easier process and lower production costs motivated the search and yielded an improved product. An ordinary polyethylene molecule was "branched," that is, its long chain of atoms possessed branches of irregular shape and length.

AT BROOKSIDE SWIM CLUB IN UNION, N.J. WHILE MOTHERS (UPPER LEFT) LOOK ON. CLUB'S ENDURANCE RECORD OF 3,000 SPINS IS HELD BY A 10-YEAR-OLD BOY

HAWKING HOOPS in Denver, public relations man Dick Fenton (in striped jacket) explains the tricks of the toy to customers at a local benefit where the hoops were demonstrated by models (left). Sales proceeds were given to charity.

HOOPING QUARRY, Wally Conrad playfully lassos date Roberta Mostal, 19, at Oak Street Beach in Chicago.

ALLEY HOOPI is performed by amiable boxer in Grosse Pointe, Mich. as Denise Renaud, 4, plays ringmaster.

"Whole Country Hoops It Up in a New Craze," September 8, 1958, excerpt from *Life* magazine

It has been estimated that a Tupperware party occurs somewhere on the planet every 2.5 seconds.

Extreme molecular length made the material strong; irregular meshing of the branched molecules made it flexible. The new linear polyethylene by contrast consisted of long, unbranched, single-chain molecules packed tightly together to create a stronger material with greater dimensional rigidity and resistance to heat.

Although the hula hoop took America by storm, the most culturally significant polyethylene product was a line of molded kitchen containers known as Tupperware, which was more responsible than anything else for domesticating thermoplastics as appropriate materials for the American home.[15] Injection molded in light, delicate colors from translucent polyethylene, Tupperware was sold from about 1950 onward only in private homes at parties given for friends and acquaintances by hostesses who hoped to earn bonus prizes by selling large quantities of the product. While Tupperware itself was one of the few inexpensive mass-marketed products to be accepted into the permanent design collection of the Museum of Modern Art, the Tupperware party became one of the most enduring symbols of the middle class in postwar America. Sidney Gross, a publicist for the plastics industry later recalled that Tupperware "gave plastics a very good name because it was terrific stuff."[16] Lightweight, tough, flexible, and unbreakable, it was equipped by its inventor Earl Tupper with a patented airtight seal that made it perfect for covered refrigerator dishes, although the line quickly expanded to include everything from mixing bowls to cocktail shakers and ice-cube trays.

Tupperware epitomized the development of plastics during the 1950s. It was versatile and nearly perfect, unless accidentally set down on a hot stove burner. And Tupperware was no mere imitation. It substituted for older materials but also offered previously unattainable qualities. Before Tupperware, people made do with storage dishes and water jugs made of ceramic and glass that were heavy and breakable; they sweated condensation and became slippery; the lids fell off and broke. Lightweight, unbreakable, and waxy textured for a good grip, Tupperware solved all those problems while also introducing the crucial innovation of the airtight seal. That it looked different simply reinforced its superiority to everything that had preceded it. Some people might have worried about a slight chemical odor, faintly suggestive of ozone, given off when an empty Tupperware container was opened, but others no doubt considered that as proof of the material's status as a product of the most up-to-date scientific research. Tupperware was one of many inexpensive products that verged on the disposable. Even its method of sale, which made it less a product than a vehicle of sociability, suggested impermanence. It proved so functional and durable, however, that people saved their Tupperware and used it for years.

Tupperware's wide social acceptance signaled the acceptance of plastics in general. Its convenience in domestic life reflected aspirations for a casual life of leisure. A sleepy householder need only watch once in disbelief as a polyethylene juice pitcher bounced off the kitchen floor without breaking to accept the superiority of new synthetic materials. Even plastic toys, despite the brittle polystyrene moldings that were already broken early on Christmas morning, proved superior in many ways. A toy soldier of molded polyethylene could not scratch the furniture nor gouge a sibling as readily as an old-fashioned tin soldier. At the same time, Tupperware's acknowledgment of plastic as plastic, as a substance with its own unique appearance and texture different from those of natural or traditional

POLYETHYLENE – 1930

Polyethylene, DDT, and Prozac
Installation view, Tang Museum

Biochemical researchers and engineers are developing biodegradable plastics that are made from such renewable resources as plants.

materials, was an idea that gathered momentum. Plastics increasingly expressed American society's fluidity and mobility, its acceptance of change for its own sake, its desire for impermanence, and its love of entertainment. Even the imitative potential of plastic, which became more convincing all the time, afforded an especially outrageous example of molding things to an arbitrary measure just because it could be done. During the last half of the twentieth century, as plastics became an ever-more-visible presence in the material world, their cultural significance was contained in the words of the skeptic John Gloag, who proposed that they be understood as what happens "when the artificial becomes the real."[17]

1. For a full historical treatment of plastics see the author's *American Plastic: A Cultural History* (New Brunswick, NJ: Rutgers University Press, 1995).

2. "What Man Has Joined Together...," *Fortune*, 13 (March 1936), 69.

3. On the history of polymerization see Yasu Furukawa, *Inventing Polymer Science: Staudinger, Carothers, and the Emergence of Macromolecular Chemistry* (Philadelphia: University of Pennsylvania Press, 1998).

4. Julian Hill, as quoted by Henry Allen, "Their Stocking Feat: Nylon at 50 and the Age of Plastic," *Washington Post* (January 13, 1988), D10.

5. For production figures see Joel Frados, *The Story of the Plastics Industry*, 13th ed. (New York: Society of the Plastics Industry, 1977), 6.

6. David M. Potter, *People of Plenty: Economic Abundance and the American Character* (Chicago: University of Chicago Press, 1954).

7. "Frontiers of Technology," *Life*, 43 (October 7, 1957), 83.

8. John Gloag, "The Influence of Plastics on Design," *Journal of the Royal Society of Arts* 91 (July 23, 1943), 466–467.

9. Thomas Hine, *Populuxe* (New York: Alfred A. Knopf, 1986), 3.

10. James W. Sullivan, interviewed by the author, October 9, 1986.

11. Eric Fawcett, as quoted by Martin Sherwood, "Polyethylene and Its Origins," *Chemistry and Industry* (March 21, 1983), 239.

12. Anonymous molder, as quoted in "1954—Fastest Growing Plastic," *Modern Plastics*, 32 (October 1954), 93.

13. Production information is from "Plastics Volume Up for 6th Year," *New York Times* (January 12, 1959), 114.

14. For 1958 production figures of hula hoops see "Hula Hoops Back in the Spin," *Modern Plastics* 45 (November 1967), 210.

15. On the history of Tupperware see Alison J. Clarke, *Tupperware: The Promise of Plastic in 1950s America* (Washington, DC: Smithsonian Institution Press, 1999).

16. Sidney Gross, interviewed by the author, December 9, 1986.

17. Gloag, "Influence of Plastics on Design" (cit. note 8), 466.

Kara Daving
CLOCKWISE FROM TOP LEFT *Aruba, Dutch Caribbean, JE (Irausquin
Boulevard)*, May 23, 2005; *Buffalo, NY (Elmwood Avenue)*,
April 19, 2005; *Bowling Green, OH (Kenwood Avenue)*, April 10, 2006;
Pittsburgh, PA (Murray Avenue), September 4, 2005
Photo transfers on plastic bags
11 x 13 inches each

NYLON 6,6

$$(C_{12}H_{22}N_2O_2)_n$$

U.S. PATENT NUMBER: 2,130,948
CAS REGISTRY NUMBER: 2,130,523

When nylon first entered the public consciousness in 1938, it claimed a novelty no other product could match. Its predecessors, rayon and cellulose, had been touted as "artificial silk," a term that implied both economy and imitation, but nylon was billed by its manufacturer, Du-Pont, as a thing unto itself. As the first commercially viable synthetic fiber, nylon ushered in a fashion revolution based on comfort, ease, and disposability. Its strength, elasticity, light weight, and resistance to mildew helped the Allies win World War II. Behind the scenes the invention of nylon transformed the chemical industry by proving that the composition of polymers could be predicted and engineered like many other chemical products. Today nylon—in toothbrushes, carpet, racket and guitar strings, surgical sutures, car parts, and, of course, hosiery—is everywhere.

E. I. Du Pont de Nemours and Company first ventured into artificial fibers in 1920 when it purchased a 60-percent interest in Comptoir des Textiles Artificiels, a French rayon company, for $4 million. The combined firm, named the DuPont Fiber Company, eventually became

by Audra J. Wolfe

the Rayon Department of the DuPont Company. Although rayon proved popular and profitable, the company expended considerable resources in improving the brittle fiber's texture and performance. In 1934 alone the company spent $1 million on rayon research.[1]

In December 1926 Charles M. A. Stine, the director of DuPont's Chemical Department, circulated a memo to the company's executive committee suggesting that it was looking in the wrong place for innovation. Rather than investing in practical research directly related to such existing products as rayon or ammonia, Stine argued, DuPont should fund "pure science work" centered on "the object of establishing or discovering new scientific facts" instead of research that "applied previously established scientific facts to practical problems." Stine's proposal was not new to industry—both General Electric and Bell Telephone operated industrial research laboratories—but his insistence that the research be "pure" or "fundamental" was fairly radical for a company focused on profits. Nevertheless, the executive committee approved a slightly modified version of the proposal in March 1927. Stine was granted $25,000 a month for research and told to hire twenty-five of the best chemists he could find. The committee also approved funds to build a new laboratory, soon dubbed "Purity Hall" by DuPont chemists.[2]

Stine encountered much more difficulty in attracting chemists to DuPont than he had anticipated, largely because academic scientists doubted whether they would truly be allowed to do "pure" research in an industrial setting. A year later, however, he made a spectacular hire when he convinced Wallace H. Carothers, a young organic chemistry lecturer at Harvard University, to join DuPont. Carothers proposed to center his research on polymerization, the process by which individual short molecules form long-chained macromolecules. Before Carothers's groundbreaking work most chemists

The DuPont company invented the name "nylon" in a process still surrounded by rumor and a degree of uncertainty. According to one apocryphal story, the name was a composite of New York (ny) and London (lon), where nylon was supposedly developed.

•

Susie Brandt
After Albers, 1995–1998
Nylon panty hose, handwoven on a potholder loom
72 x 58³/₄ inches

based their polymers on complicated "recipes" largely determined by chance. Moreover, the nature of polymers was poorly understood, with some researchers convinced that the sticky resins represented complex colloidal systems while others advocated the long-chain molecule theory originally advanced by Hermann Staudinger, a German chemist. Carothers hoped to offer definitive proof of Staudinger's theory by constructing polymers from small organic molecules with known reactivity at both ends.[3]

Carothers's success was almost immediate. In April 1930 Julian W. Hill, a research associate in Carothers's group, created a polyester with a molecular weight of over 12,000 by combining a dialcohol and a diacid to produce a long polymeric ester—the first "polyester." Hill's polyester fibers had a remarkable property: when cooled, the thin, brittle filaments could be pulled into an elastic thread four times their original length. DuPont researchers soon realized, however, that this first polyester would never succeed as a commercial fiber because its low melting point made laundering and ironing impractical.

Nylon is six times stronger than steel of the same weight.

For the next four years attempts to create commercially viable synthetic fibers were stymied by the twin problems of low melting points and high solubility in water. In 1934 Elmer Bolton, the new chemical director at DuPont, urged Carothers to return to the problem. Carothers agreed, but this time he would focus on polyamides rather than polyesters. On May 24, 1934, Donald D. Coffman successfully pulled a fiber of a polymer based on an aminoethylester. His fiber—ultimately the first nylon—retained the remarkable elastic properties of the polyesters but lacked their drawbacks. However,

ABOVE Wallace Carothers with neoprene, 1930
OPPOSITE Julian Hill reenacting the 1930 discovery of cold drawing, 1948

since the intermediate used to form the polymer, aminononanoic ester, was tremendously difficult to produce, Carothers and his associates kept looking.

Within a year Carothers's six researchers had narrowed the field to two possibilities: polyamide 5,10, made from pentamethylene diamine and sebacic acid; and polyamide 6,6, made from hexamethylene diamine and adipic acid. (The molecules are named for the number of carbons in the starting materials.) Carothers preferred 5,10, but Bolton pushed for 6,6 since the intermediates could be more easily prepared from benzene, a readily available starting material derived from coal tar. As Carothers's declining mental health kept him increasingly absent from the laboratory, Bolton's choice triumphed, and all hands turned to improving fiber 6,6.[4]

Joseph Labovsky, a chemical engineer working as a technician in the lab, later recalled that the lab workers were scaling up fiber 6,6 "from one ounce to one pound, two pounds, fifty pounds, two hundred fifty pounds, and eventually to two thousand pounds."[5] Paul Flory, a young physical chemist who would later win the Nobel Prize in Chemistry for his work on polymers, helped the researchers stabilize the reaction by developing a mathematical model for the kinetics of the polymerization reaction.[6] In 1938 DuPont started construction on a nylon production facility in Seaford, Delaware, that could produce up to twelve million pounds of the synthetic fiber a year. It was time to introduce nylon to the American public.[7]

Nylon's characteristics made it an ideal material for any number of uses, but DuPont decided early on that it would focus on a single market: ladies "full-fashioned" hosiery. As hemlines continued

From 1939 to 1962 nylon production went from fewer than fifty thousand pounds per year to over 600 million pounds per year and was DuPont's most profitable product.

Nylon
Installation view, Tang Museum

USES
FAR
DE

The continued vitality of nylon as a commercial product rests upon a quality: it has demonstrated throughout its war history — its ability to prove useful in so many walks of life, ranging from the very start, and every new and promising find a foothold in a new application.

Fire yarn is especially revealing of nylon's utility. Although it costs more per pound than rayon, its greater strength has made nylon the fiber in over 70 per cent of replacement passenger tires, and in over 72 per cent of replacement tires for trucks and buses. Meanwhile, for the improved nylon tire cords are being made. A new low-flat-spotting type may soon make heavy inroads in the original equipment market long dominated by rayon.

Nylon has also found a wide variety of uses in recreational materials. Inside an elastomeric shell, it keeps basketballs round. In tenting and other mountainclimbing gear, it has helped keep camps alive on the pinnacles of Mt. Everest. In ski pants, in stretch pants, it keeps millions of it popular it keeps inches or at glamour roats snug in berths or at mooring. Nylon netting devices snag golf balls for backyard or basement practice All glider tow ropes are made of nylon.

SAILCLOTH IS USED IN SPINNAKERS (LARGE BILLOWING SAIL USED TO CATCH A FOLLOWING WIND) HAS BEEN A POPULAR FIBER THE SEAS AROUND SINCE IT WAS INTRODUCED.

NYLON FOR TIR...

NYLON FOR CARPETS now totals over 150 million pounds a year. Nylon is easily cleaned, evenly bright color, has good resilience. New translucent type used in carpets has helped nylon take over 26 per cent of market for broadloom, carpeting.

NYLON FOR FIREARMS replaces traditional plastic hand stocks with engineered durable plastic that is light, tough, Remington's XP-100 target-varmint pistol uses a single joint stock and grip molded of Zytel nylon resins. Nylon stocks have also proved popular in U.S. glamorous line of Remington in 22. Another popular use of nylon is in "Stren" sparkling monofilament fishing lines.

N THE U.S.
SHARP

...ns mean much to nylon plant town...
...First National Bank, Martinsville, Va.

...na as quickly
...e San Fran...
...tion. Near...
...n war it
...ses

23

to rise throughout the 1930s, silk and rayon stockings had become an increasingly necessary part of every woman's wardrobe. American women bought an average of eight pairs of stockings per year, earning Japanese silk producers over $70 million. DuPont never intended to produce the stockings directly; rather, the company would provide nylon thread to mills that would knit and sell the hosiery.[8]

Before DuPont could take its new miracle fiber to the public, however, its leaders had to decide what to call it. In-house researchers had alternately been referring to what would become nylon as Rayon 66, Fiber 66, or "Duparon," a creative acronym for "DuPont pulls a rabbit out [of] nitrogen/nature/nozzle/naptha." In 1938, through a decision-making process that remains somewhat obscure, the company

settled on the word *nylon.* According to Ernest Gladding, manager of the Nylon Division in 1941, the name had originally been *Nuron,* which not only implied novelty but conveniently spelled "no run" backwards. Unfortunately, *nuron* and other closely related words posed trademark conflicts, so the division proposed *nilon.* Changing the *i* to a *y* removed any ambiguity surrounding pronunciation, and "nylon" was born. The company then decided not to trademark the name, hoping instead to encourage consumers to think of nylon as a generic preexisting thing, like wood or glass.[9]

Since 1931, when Carothers first reported on his polyester fibers at an American Chemical Society meeting, newspapers had been reporting rumors that DuPont had developed a new fiber as good as or better than silk. By early 1938 the press was producing a steady stream of articles suggesting that stockings made from the mystery fiber would outlast silk and never run. If DuPont executives were beginning to grow nervous about unrealistic expectations, they grew truly alarmed in September 1938 when the *Washington News* ran a story based on the newly released patent (U.S. 2,130,948). The article claimed that nylon could be prepared from cadaverine, a substance formed during putrefaction in dead bodies. When combined with reports of Carothers's suicide earlier that year, coverage of nylon took on an oddly morbid tone. Perhaps to counteract these rumors, for many years thereafter DuPont's publicity department stressed that nylon was derived solely from coal, air, and water.[10]

DuPont regained control of nylon's publicity on October 27, 1938, when it officially introduced the stockings to a crowd of four thousand enthusiastic middle-class women at the future site of the New York World's Fair. But while the excitement was building, the stockings themselves would not become commercially available for another eighteen months. At that point the only women who could experience the stockings firsthand either worked for DuPont or were married to DuPont scientists in the nylon division. A limited supply of the first pairs went on sale in Wilmington, Delaware, in October 1939, but the stockings did not finally reach the national market until May 15, 1940. Offered at $1.15 a pair, they were sold out at most locations by noon. In 1940 DuPont produced 2.6 million pounds of nylon for total sales of $9 million; the following year the company sold $25 million worth of nylon yarn. Within two years of nylon's introduction DuPont had captured an astonishing 30 percent of the full-fashioned hosiery market.[11]

Sixty-four million pairs of nylons were sold in 1940, the first full year of production. The first four million pairs were sold in four days!

OPPOSITE Silver gelatin photo, ca. 1950

Alas, American women's access to nylon hosiery proved short-lived. In November 1941 Du-Pont shifted its nylon production from consumer to military production as a replacement for Japanese silk. In 1940, 90 percent of DuPont's nylon went into stockings; by 1942 virtually all nylon went into parachutes and tire cords. Nylon would eventually be used in glider tow ropes, aircraft fuel tanks, flak jackets, shoelaces, mosquito netting, and hammocks. Inevitably, nylon found its way onto the black market, with one entrepreneur making a hundred thousand dollars off of stockings produced from a diverted nylon shipment.[12]

DuPont jumped back into consumer nylon production almost as soon as the war ended, with the first pairs of stockings arriving in stores in September 1945. Everywhere the stockings appeared, newspapers reported on "nylon riots" in which hundreds, sometimes thousands, of women lined up to compete for a limited supply of hosiery. Perhaps the most extreme instance occurred in Pittsburgh in June 1946, when forty thousand people lined up for over a mile to compete for thirteen thousand pairs of nylons.[13] Labovsky recalled that demand remained so high throughout the 1940s that DuPont required all its customers, no matter how large or reputable the account, to pay in advance. "The demand was *so great*. We had to make sure customers who wanted nylon had the money to pay for it.... Even Burlington

ABOVE Betty Grable auctioning her nylons at a
war bond rally, 1943
OPPOSITE DuPont nylon dress advertisement, *McCall's*,
November 1956

Mills would send a check for a hundred thousand dollars to fill an order.... Everybody wanted nylon."[14] Partially to meet demand and partially to avoid an antitrust suit, DuPont finally licensed nylon to outside producers in 1951.[15]

Nylon stockings were only the beginning of what would soon become a fashion revolution. Cheap and colorful, synthetic fibers offered the promise of an easy-care, wash-and-wear, disposable future. By the 1950s nylon and other synthetic fibers could be found in underwear, socks, petticoats, fake fur coats, mock-wool sweater sets, and even men's drip-dry suits. Women's fashion was especially transformed by synthetic fabrics, as new Lycra girdles—more comfortable and lightweight than traditional rubber models—cinched women's bodies into dramatic hourglass figures that could then be surrounded with yards and yards of billowing synthetic material.

Because the variety of synthetic fibers was basically limited to viscose (rayon), acetates, polyesters, and polyamides, manufacturers realized early on that the key to their success lay in branding their specific products as unique. Generic DuPont nylon was soon joined in the marketplace by Bri-Nylon, Dacron (polyester), Terylene (polyester), Crimplene (polyester), Orlon (acrylic), Acrilan (acrylic), Tricel (acetate), and seemingly dozens more. Each of the chemical companies making these products then launched extensive advertising campaigns aimed at winning consumers' loyalty to a specific fabric rather than to the specific fashions of a given season.

DuPont developed a particularly sophisticated approach to marketing its synthetic fibers.

At the time of nylon's development owning nylon stockings was a status symbol; some women even painted dark lines on the backs of their legs to simulate stocking seams.

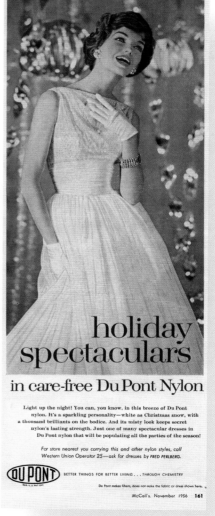

holiday spectaculars
in care-free Du Pont Nylon

Light up the night! You can, you know, in this breeze of Du Pont nylon. It's a sparkling personality—white as Christmas snow, with a thousand brilliants on the bodice. And its misty look keeps secret nylon's lasting strength. Just one of many spectacular dresses in Du Pont nylon that will be populating all the parties of the season!

For store nearest you carrying this and other nylon styles, call Western Union Operator 25—ask for dresses by FRED PERLBERG.

DU PONT BETTER THINGS FOR BETTER LIVING ... THROUGH CHEMISTRY
REG. U.S. PAT OFF.

De Pont makes fibers, does not make the fabric or dress shown here.

McCall's, November 1956 **161**

Nylon parts
in automobiles are
substitutes for
metal, making vehicles
lighter and thus
increasing gas mileage.

From the earliest days of its rayon production DuPont realized that if it were to capture the textile market, it needed to capture the hearts of Parisian couturiers. The company's Fabric Development Department, established in 1926, worked with designers to produce sample fabrics for textile mills and clothing manufacturers. By the mid-1950s the group was producing well over a thousand fabric samples each year. The DuPont sales force then attempted to sway fashion designers by providing them with generous samples and free publicity. Their first dramatic success occurred at the 1955 Paris fashion shows, in which at least fourteen synthetics featuring DuPont fibers appeared in gowns from Coco Chanel, Jean Patou, and Christian Dior. To heighten the glamour DuPont recruited fashion photographer Horst P. Horst to document designers' works and then circulated the photographs in press releases across the country. Besides couture by Chanel, Dior, and Patou, Horst's photos featured gowns by Madame Grès, Maggie Rouff, Lavin-Castillo, Nina Ricci, Emmanuel Ungaro, Philippe Venet, Pierre Cardin, and the New York Couture Group, all in DuPont fabrics. A decade later, vanguard 1960s designers Pierre Cardin and André Courrèges embraced the futuristic feel of synthetics as the right look for space-age living.[16]

By the late 1960s synthetics had moved firmly off the runways and into the mass markets—and therein lay their downfall. Victims of overexposure, nylon and polyester seemed suddenly out of date, and their shiny luster started to look tacky. In the wake of Rachel Carson's *Silent Spring* (1962) and a growing environmental movement consumers were turning to natural fibers, particularly cotton and wool. In 1965 synthetic fibers made up 63 percent of the world's production of textiles; by the early 1970s that number had dropped to 45 percent. Although synthetic fibers regained some of their popularity in the 1990s as technical innovations improved their feel and performance, never again would synthetic fibers dominate the market as they did in the 1950s and 1960s.[17]

Yet nylon is here to stay. We may not be wearing it as much, but in one form or another nylon surrounds us in our homes, offices, leisure activities, and transportation. The polymer revolution ushered in by nylon's discovery has left us with a world of plastics that would be unrecognizable to our grandparents' generation. Today manufacturers worldwide produce around eight million pounds of nylon, accounting for about 12 percent of all synthetic fibers. Nylon may no longer be DuPont's most profitable product, but it remains one of its most important inventions.

1. David A. Hounshell and John Kenly Smith, Jr., *Science and Corporate Strategy: Du Pont R&D, 1902–1980* (New York: Cambridge University Press, 1988), 161–169.

2. Ibid., 223–227.

3. For concise explanations of both Carothers's decision to accept DuPont's offer and his understanding of polymerization, see Jeffrey L. Meikle, *American Plastic: A Cultural History* (New Brunswick, NJ: Rutgers University Press, 1995), 129–131; Hounshell and Smith, *Science and Corporate Strategy* (cit. note 1), 229–232; and Pap A. Ndiaye, *Nylon and Bombs: DuPont and the March of Modern America*, trans. Elborg Foster (Baltimore: Johns Hopkins University Press, 2007), 93. For a detailed biography of Carothers see Matthew E. Hermes, *Enough for One Lifetime: Wallace Carothers, Inventor of Nylon* (New York: American Chemical Society and the Chemical Heritage Foundation, 1996). An excellent study of the debate over the nature of polymers is found in Yasu Furukawa, *Inventing Polymer Science: Staudinger, Carothers, and the Emergence of Macromolecular Chemistry* (Philadelphia: University of Pennsylvania Press, 1998).

4. The story of various chemical combinations tried by Carothers and his associates is recounted in Hounshell and Smith, *Science and Corporate Strategy* (cit. note 1), 236–262; and Meikle, *American Plastic* (cit. note 3), 130–137. The appendix in Hermes, *Enough for One Lifetime* (cit. note 3), explains many of the chemical reactions.

5. Joseph Labovsky, interview by John K. Smith, July 24, 1996, bound volume with subsequent corrections and additions, Oral History Collection, Chemical Heritage Foundation, Philadelphia, p. 28.

6. Peter J. T. Morris, *Polymer Pioneers: A Popular History of the Science and Technology of Large Molecules* (Philadelphia: Beckman Center for the History of Chemistry, 1990), 70–73.

7. Hounshell and Smith, *Science and Corporate Strategy* (cit. note 1), 268. For more details on the production development process see Labovsky interview (cit. note 5).

8. Meikle, *American Plastic* (cit. note 3), 137; and Hounshell and Smith, *Science and Corporate Strategy* (cit. note 1), 257.

9. Meikle, *American Plastic* (cit. note 3), 137–138. See also "About DuPont Nylon," a booklet produced by DuPont's Nylon Division (1946), copy available at the Othmer Library of Chemical History, Chemical Heritage Foundation, Philadelphia.

10. Meikle, *American Plastic* (cit. note 3), 139–140.

11. Ibid., 141–146; and Hounshell and Smith, *Science and Corporate Strategy* (cit. note 1), 266–273. Note that Hounshell and Smith put the number of women at the initial announcement at three thousand rather than four thousand.

12. Meikle, *American Plastic* (cit. note 3), 148.

13. Ibid., 149–150.

14. Labovsky interview (cit. note 5), p. 42.

15. Meikle, *American Plastic* (cit. note 3), 151.

16. This account is drawn largely from Susannah Handley, *Nylon: The Story of a Fashion Revolution* (Baltimore: Johns Hopkins University Press, 1999), which is by far the best source on synthetic fibers and the fashion industry.

17. Ndaiye, *Nylon and Bombs* (cit. note 3), 222–226.

Nylon
Installation view, Tang Museum

1950

DNA

U.S. PATENT NUMBER: 5,744,305
CAS REGISTRY NUMBER: 9007-49-2

In 1953, James Watson and Francis Crick, working together at the University of Cambridge in England, accomplished a major breakthrough in science. They proposed the double-helical structure for DNA, one of the long-pondered mysteries of science. In an article published in *Nature* on April 25, 1953, they discussed their findings: "We wish to suggest a structure for the salt of Deoxyribose Nucleic Acid (DNA). This structure has novel features which are of considerable biological interest."[1] We now know that these modest words were the drastic scientific understatement of the century! The discovery of the double helix has not only led us to advances in the emerging fields of biochemistry, molecular biology, and genetics but has also provided us with a new understanding of diseases and novel ways to cure them. It has created new avenues in the practical applications of science through biotechnology, genetic engineering, and forensics. DNA has become a household word in our society: browse through any magazine or newspaper on any day, and you are certain to find stories and articles involving DNA; watch any television show that deals with law, medicine, crime, or investigation, and you will find ample references to DNA. The term pops up routinely even in commercials for health and wellness products.

by Vasantha Narasimhan

But what is DNA? What is its history? What is the reason for its claim to fame as the "secret of life"?

Ever since the dawn of intellectual curiosity humans have pondered the origin of heredity as they observed that close relatives tended to be similar in their physical features and personality traits. Though our ancestors had no answer to this riddle, they used the observation to their benefit through careful selection and breeding techniques for animals and plants, which were customized for human use. The principles of heredity were understood by ancient farmers long before nineteenth-century scientists began exploring the underlying theories.

Gregor Mendel published his famous paper on the theory of heredity in 1866.[2] He proposed that specific factors (later called genes) passed from parent to offspring. Working with green (G) and yellow (Y) peas, he theorized that heredity factors came in pairs, and the offspring received one factor from each parent. For example, if the offspring received one G factor from each parent, the resulting pea (GG) would be colored green. If it received a Y factor from each parent, the resulting pea (YY) would be yellow. If it received a G from one parent and a Y from the other (GY), depending on which was the dominant factor and which the recessive factor, the offspring pea would be either yellow or green. He was far ahead of his time with his careful experimentation, keen observation, and expert analysis. But his results were buried in an obscure journal and were unintelligible to most biologists of his era. Moreover, they contradicted the existing theories about the origin of life and implied that actual "things" or "material objects" were transmitted from generation to generation. The rediscovery of Mendel's work thirty-four years later by three plant geneticists was instrumental to the recognition of the value of that work by the scientific community. In 1909 William Bateson, a British biologist, gave the name "genetics" to the science of inheritance.[3]

DNA is a natural polymer, a chain of repeating units; it is a very thin molecular wire.

Alexis Rockman
Romantic Attachments, 2007
Oil and wax on wood panel
120 x 96 inches

The discovery of DNA gives us a deep foundational comprehension of our biological identity as well as our physical connectivity to other species.

While Mendel was working to understand the origin of heredity through his pea-plant studies, Friedrich Miescher, a German chemist, took on the task of analyzing the chemical composition of human blood cells. He gathered large quantities of pus from discarded bandages and isolated a new molecule from the dead white blood cells. It was a white, fibrous, acidic substance rich in nitrogen and phosphorus along with carbon, hydrogen, and oxygen. In Miescher's paper published in 1871 he named the substance "nuclein";[4] it wasn't until much later that it was renamed deoxyribonucleic acid (DNA). Though Mendel and Miescher were contemporaries, no one speculated that Miescher's compound was the key to explain Mendel's heredity theory. Little did the scientific world know then that these two discoveries were the foundation for the eternal marriage of two branches of science: biology and chemistry.

Although the chemistry of DNA was studied by several investigators after its discovery by Miescher in 1866, three-quarters of a century elapsed before its biological significance was realized. The reason for this hiatus was the belief that proteins were the major determinants of all life processes. It was strongly believed that the nuclein isolated by Miescher was nothing but a reaction product of protein and phosphoric acid that were primary components of cellular fluids. In 1944 three bacteriologists from the Rockefeller Institute—Oswald Avery, Colin McLeod, and Maclyn McCarty—wrote a paper published in the *Journal of Experimental Medicine* that later became a classic.[5] In the article they identified DNA as the transforming principle. They worked on transforming pneumonia bacteria from virulent to benign species by systematically eliminating either the DNA or the protein components of the bacteria. They found that, if DNA was destroyed, the transformation did not occur, but destroying the protein did not affect transformation. Although the three met with protests from protein enthusiasts, they managed to move DNA into the genetic limelight through these studies.

Following the revelation that DNA rather than protein was the genetic information carrier, the structure of DNA became a hot topic for research and a subject of intense competition among researchers for the next decade. This "race for the double helix" continued, with physicists, chemists, biologists, and medical researchers working on various aspects of the double-helix structure. The race was so hotly contested in scientific circles that bets were made on probable winners. In fact, Linus Pauling, the supreme patriarch of structural chemistry, who had successfully solved the alpha-helical structure of protein around the same time, was the top choice of many.[6]

In early 1951 Erwin Chargaff, an Austrian-born chemist working at the Columbia University College of Physicians and Surgeons, measured the relative amounts of the four bases—adenine (A), guanine (G), cytosine (C), and thymine (T)—in several DNA samples using a new technique called paper chromatography.[7] He found that in any sample of DNA the percentage of A was always equal to that of T and the percentage of G was always equal to that of C. Further, the relative percentage of the four bases (base composition) was always the same in the DNA samples isolated from different cells of the same animal. He also discovered that DNA samples from different species had different base compositions. Though all these facts may seem to complicate things, they actually confirmed that each species, whether animal, plant, or bacteria, is characterized by its unique DNA composition. In other words, by chemically analyzing a sample of DNA, one can tell from which

The DNA of chimpanzees is 98 percent identical to that of humans; chimps are our closest living relatives.

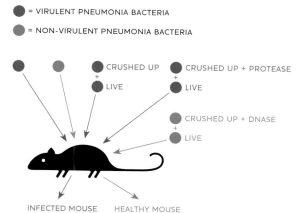

● = VIRULENT PNEUMONIA BACTERIA

● = NON-VIRULENT PNEUMONIA BACTERIA

CRUSHED UP
+
LIVE

CRUSHED UP + PROTEASE
+
LIVE

CRUSHED UP + DNASE
+
LIVE

INFECTED MOUSE   HEALTHY MOUSE

A schematic view of Avery's bacterial transformation experiment

Bryan Crockett
From the *Seven Deadly Sins* series, 2001
CLOCKWISE FROM TOP
Anger, Gluttony, Sloth

species it came. This principle has indeed led to the much-talked-about "DNA test" used in forensics and paternity checks during the last two decades.

Two other significant events took place while Chargaff was working out the chemistry of DNA. The first was the publication of *What Is Life?* by Erwin Schrodinger, a great physicist of that time. Schrodinger argued that life could be thought of in terms of storage and transmission of biological information from generation to generation in the form of chemical units. Since a lot of information had to be packed in every cell, Schrodinger proposed that it must be coded similarly to the Morse code of dots and dashes and embedded in the tiny chromosomes. He suggested that identifying these molecules and cracking the code would lead us to solving the mystery of life and the origin of heredity. Watson, Crick, and Maurice Wilkins, three key participants in the race for the double helix, were tremendously influenced by Schrodinger's ideas.

The second significant event was the X-ray crystallography studies of DNA samples by Wilkins, an English physicist, and his coworker Rosalind Franklin. Wilkins had been involved in the Manhattan Project during World War II. The actual deployment of the bomb on Hiroshima and Nagasaki disillusioned him to the extent that he wanted to leave science altogether and become a painter in Paris. But Schrodinger's book caught his attention, and he shifted his research focus toward solving the DNA puzzle. At King's College in London, Wilkins and Franklin both worked on solving the crystal structure of DNA. Franklin finally succeeded in obtaining a clear and simple X-ray crystal pattern of DNA, which is commonly referred to as "Photo 51."

Human cells contain forty-six chromosomes, with twenty-three inherited from each parent.

Molecular structures of the four base pairs of DNA and deoxyribose phosphate

DNA
Installation view, Tang Museum

Watson and Crick orchestrated the grand finale of the DNA story as they worked together at Cambridge University. James Watson, who received his Ph.D in zoology from Indiana University, had worked on understanding the genetics of phages (virus-like particles that infect bacteria). From his studies he could come only to biological conclusions regarding the mode of action of the phages. He realized that getting complete answers required an understanding of the intricate chemical details of DNA contained in the phages. Impressed by Schrodinger's idea of coded genetic information, Watson decided to study the chemical nature of DNA by joining a research group in Copenhagen that was pursuing the chemical analysis of DNA and the synthesis of some of its individual components. During that time he

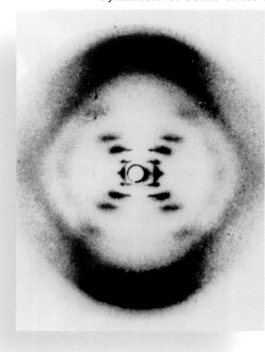

attended a conference in Naples on X-ray crystallography of biomolecules and was fascinated by the type of structural information that this technique could yield. Watson met Wilkins at the conference and learned of his X-ray studies on DNA. Soon after the conference Watson joined the Cavendish Laboratory at Cambridge University in London where he met Francis Crick. Crick was an ex-physicist who had worked on military research related to magnetic mines during World War II. As he read Schrodinger's book *What Is Life?* his fascination with the intricacies of scientific inquiry and his enormous intellectual curiosity were sparked. His interest turned toward biology and the chemistry of DNA. Together Watson and Crick embarked on

X-ray crystallographic pattern of DNA
obtained by Rosalind Franklin (Photo 51), 1951

their mission to solve the molecular structure of DNA by building various models of the molecule that would fit the biological, chemical, and genetic data on DNA documented in the literature. Wilkins, who met Watson and Crick during their visit to King's College, showed them Franklin's yet unpublished X-ray diffraction picture of DNA (Photo 51). It became clear to Watson and Crick that the pattern in Franklin's picture could only result from an ordered helical structure.

Meanwhile, Linus Pauling, who had just accomplished the amazing feat of deducing the alpha-helical structure of proteins, put forth suggestions that the gene probably consisted of mutually complementary strands of DNA perhaps intertwined into a triple helix. Because of Pauling's involvement in the peace movement and the "Ban the Bomb" campaign, he was denied a passport to attend a crucial conference on proteins held in England in 1952. Some scientists speculated that if Pauling had attended that conference, he would have had the opportunity to meet Franklin and would have seen her X-ray data on DNA. Then he rather than Watson and Crick might have deduced the double-helix structure of DNA.

Franklin's Photo 51, Chargaff's rules on DNA base composition, and Avery's transformation data were a few of the key factors that led Watson and Crick to hypothesize a double-helical structure for DNA. They constructed several helical models with cardboard cutouts and handmade wooden models of atoms. Finally, there it was, on February 28, 1953: a six-foot-tall model containing two helical threads of DNA running opposite to each other and wound around in a double helix. The model adhered to Chargaff's rules on base composition of DNA. Base-stacking and base-pairing hypotheses now made sense, the role of DNA as the repository of genetic codes was possible, the X-ray data fit perfectly, and even the origin of genetic mutations could be explained.

The linear length of all the DNA in any cell in the human body is approximately six feet. With an estimated one hundred trillion cells there is enough DNA in one person to span the distance to the sun and back about five hundred times.

Michael Oatman
Code of Arms, 2004
Collage on
printed paper with
test-tube rack frame
106 x 54 inches

In their modestly titled 1953 paper "A Structure of the Deoxyribose Nucleic Acid," Watson and Crick describe the DNA double helix as

> two helical strands coiled around the same axis. The two chains are related by a dyad (the pairing of opposite bases) perpendicular to the fibre axis. Both chains follow right handed helices, but owing to the dyad the sequences of atoms in the chains run in opposite directions (anti-parallel). The bases are on the inside of the helix and the phosphates on the outside.... It has not escaped our notice that the specific pairing we have postulated immediately suggests a possible copying mechanism.[8]

Of course, they were right in calling their model "a" structure and not "the" structure. With the advent of advanced computational technologies and refinements in experimental methods and instrumentation, scientists now know that at least twenty valid double-helix models can be visualized for any given DNA. As a matter of fact two other DNA structures, the A and Z double-helical structures, have been detected in sections of DNA in actual living cells. The original Watson-Crick double helix is now called the B structure of DNA. Watson, Crick, and Wilkins were awarded the Nobel Prize in Physiology or Medicine in April 1962 for their landmark discoveries "concerning the molecular structure of nucleic acids and its significance for information transfer in living material."[9]

In the decades since the discovery of the double helix a lot has happened in terms of the scientific, technological, and societal applications of DNA and its role as a genetic determinant. The

James Watson (left) and Francis Crick (right), seen with their model of part of a DNA molecule in 1953

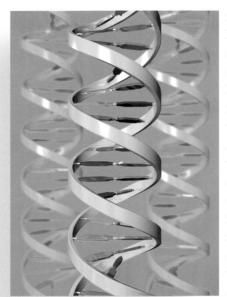

manipulation of DNA within genes has resulted in several breakthroughs in medicine, genetics, agriculture, forensics, and anthropology. Conditions ranging from obesity to severe acute respiratory syndrome, or SARS (a deadly flu-like illness), have been tackled through gene sequencing, modification, and repair. Researchers have experimented with DNA-based, noninvasive diagnostic tests for cancer. A research group has recently reported a DNA-based blood test that may help predict the risk for developing colon cancer.[10] Genetically modified crops and foods are produced by manipulating the genes of crops to yield more nutritious, safer foods with a longer shelf life for the growing world population. With the refined technological tools available today anthropologists have been able to analyze the mitochondrial DNA from Neanderthal fossils and have concluded that Neanderthals and modern humans had a common ancestor about five hundred thousand years ago.[11] Refinements of a molecular copying technique called polymerase chain reaction have enabled scientists to analyze minute amounts of DNA (the so-called DNA test or DNA fingerprinting) for forensic investigations. This test has not only helped solve crimes and settle paternity suits; it has also helped exonerate wrongfully convicted persons.[12]

In 1990 the Human Genome Project was launched by an international group of scientists, and Watson was appointed its director by the National Institutes of Health. After years of effort the sequencing of the human genome was completed, and now scientists have the "blueprint of life" in their hands. It is hoped that knowledge of the complete sequence of the human genome may help scientists discover the genetic origins of such diseases as Alzheimer's, cystic fibrosis, several types of cancer, heart disease, diabetes, and obesity. Within

DNA double helices

the next decade or two, they may be able to offer sophisticated tools to health professionals for the prevention and cure of these and many other diseases. Of course, some of these new explorations that tamper with the DNA molecule in our very genes have met with skepticism and opposition from the public, but the immense boons that the new discoveries offer society seem to outweigh the possible harms as we strive to achieve health, long life, and a safe environment.

1. J. D. Watson and F. H. C. Crick, "A Structure of the Deoxyribose Nucleic Acid," *Nature* 171 (1953), 740–741.

2. Gregor Mendel, "Versuche über Pflanzenhybriden" [Experiments on plant hybridization], *Verhandlungen des naturforschenden Vereines in Brünn* 4 (1866), 3–47.

3. James D. Watson and Andrew Berry, *DNA: The Secret of Life* (New York: Alfred A. Knopf, 2003), 15.

4. Friedrich Miescher, "Ueber die chemische Zusammensetzung der Eiterzellen" [On the chemical composition of pus cells], *Medizinisch-chemische Untersuchungen* 4 (1871), 441–460.

5. O. T. Avery, C. M. McLeod, and M. McCarty, "Studies on the Chemical Nature of the Substance Inducing Transformation of Pneumococcal Types: Induction of Transformation by a Deoxyribonucleic Acid Fraction Isolated from Pneumococcus Type III," *Journal of Experimental Medicine* 79:2 (1944), 137–158.

6. Valley Library, Oregon State University, "Linus Pauling and the Race for DNA," osulibrary.orst.edu /specialcollections/coll /pauling/dna/index.html (accessed July 2, 2007).

7. E. Chargaff et al., "The Composition of Desoxypentose Nucleic Acids of Thymus and Spleen," *Journal of Biological Chemistry* 177 (1949), 405–416.

8. Watson and Crick, "A Structure" (cit. note 1).

9. Nobel Foundation, "The Nobel Prize in Physiology or Medicine 1962," nobelprize.org /nobel_prizes/medicine /laureates/1962/ (accessed July 2, 2007).

10. H. Cui et al., "Loss of IGF2 Imprinting: A Potential Marker of Colorectal Cancer Risk," *Science* 299 (2003), 1,753–1,755.

11. R. G. Klein, "Whither the Neanderthals?" *Science* 299 (2003), 1,525–1,527.

12. Tim Junkin, *Bloodsworth: The True Story of One Man's Triumph over Injustice* (Chapel Hill, NC: Algonquin Books, 2004).

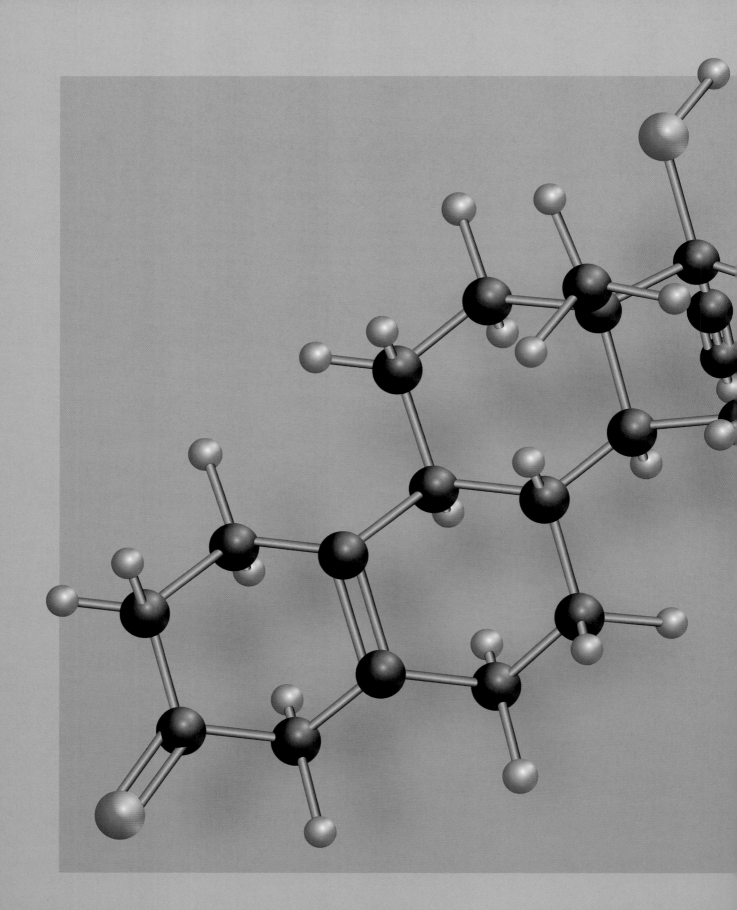

1960

PROGESTIN

$$C_{20}H_{26}O_2$$

U.S. PATENT NUMBER: 3,409,721
CAS REGISTRY NUMBER: 57-83-0

In the history of twentieth-century medicine only one substance has achieved a status so renowned, so controversial, and so familiar that we know it neither by its biological function nor by its chemical content, but simply as "the Pill." Whether as an icon of women's sexual liberation, the cultural upheavals of the 1960s, or medical progress, the Pill stands alone. The molecules behind this drug are called progestins, synthetic variants of the natural hormone progesterone. By suppressing ovulation progestins have allowed sexually active women to control their fertility more effectively in recent years than at any other time in human history.

The story of the Pill begins with the surprisingly late discovery in the early years of the twentieth century of the actual mechanisms of human ovulation and fertilization. Scientists in Scotland, France, and Switzerland studied animal reproduction and observed that pregnant animals did not ovulate; eventually they identified the *corpus luteum*, the remains of the egg sac in the ovary, as containing a substance that appeared to prevent ovulation.[1]

by Mary C. Lynn

The tale of the Pill must also note the subsequent discovery of steroid hormones, chemical messengers that regulate ovulation, conception, and pregnancy, and the isolation of two such hormones, progesterone and estrogen. (Progesterone is produced by the *corpus luteum* and eventually by the placenta and prevents further ovulation during pregnancy.) Natural steroid hormones were expensive, so much so that their discovery was followed by a rush to create synthetic versions. This activity involved the contributions of the brilliant but eccentric "fathers of the Pill"—the chemists who discovered progestin and the physicians who administered it to women, first to solve problems of infertility and only later to regulate fertility.

Russell E. Marker, whose name rarely appears in reference sources without the adjective *eccentric* appended, synthesized progesterone from the roots of a Mexican wild yam in 1943. He later started Laboratorios Syntex S.A. (a name derived from combining *synthesis* with *Mexico*) to develop the product. In 1951 Carl Djerassi, a chemist working at Syntex, developed norethindrone, a progestin that could be taken orally; in contrast progesterone had to be administered by injection.[2] Other researchers were working to create new steroid drugs during this period, including Frank Colton, a chemist at the drug company G. D. Searle, who developed another synthetic progestin, norethynodrel, patented in 1953. The umbrella term *progestin*, abbreviated from *progestational,* was applied to these and other products that at the time were being developed as treatments for infertility or menstrual problems, not as contraceptives.[3] Gregory Pincus, a biologist at Harvard University, studied in vitro fertilization of rabbits in the 1930s until he lost his bid for tenure, perhaps owing to public criticism of such "Frankenstein" work. He found refuge at the tiny Worcester Foundation for Experimental Biology, where he, too, carried out research on steroids. He and his colleague Min-Chueh Chang began to use Marker's

Progestin functions biologically as progesterone, the naturally occurring female hormone. By mimicking progesterone, progestin prevents ovulation and therefore conception.

yam-derived progesterone first in rabbits and then in rats to prove that it
would suppress ovulation. Ironically,
the Worcester Foundation was located
in Massachusetts, a state that in the
1870s had outlawed not only contraceptive devices but even the provision of
contraceptive information, laws that remained on the books, enforced sporadi-
cally, until 1972.[4]

Birth-control pills
were the first
prescription drug
designated for use by
healthy women.

Pincus went on to supply progestin to John Rock, a Harvard gyne-
cologist and infertility specialist who was exploring whether the
chemical suppression of ovulation in infertile women would create a
"rebound effect" when the progestin was stopped, thereby jump-
starting ovulation and making pregnancy possible. Despite some success with this treatment Rock knew that the progestin he used was
indeed a possible contraceptive.[5] He proved an especially useful ally
to the contraceptive cause since his professional stature was earned
as an infertility expert; he was, however, also concerned about over-
population and sympathetic to the search for a scientific solution.[6]

If these scientists were the fathers of the Pill, then two feminist
birth-control advocates were its mothers. Margaret Sanger and Katharine Dexter McCormick met in 1917 and began working together in
1923. McCormick, long active in the fight for suffrage, was searching
for ways to consolidate women's 1920 victory in the passage of the
Nineteenth Amendment to the Constitution, which gave women the
right to vote.

Sanger's story is well known—a daughter who saw her mother
dying young, worn out by childbearing, and who became a nurse

ABOVE LEFT Gregory Pincus, 1962
ABOVE RIGHT John Rock, 1961

and worked with poor immigrants in the slums of New York. She became determined to offer poor women the knowledge about birth control and access to it already enjoyed by richer and better-informed women. Opening her first clinic led to Sanger's arrest, but she was

uncowed. By the early 1920s she had built alliances with elite women and with physicians, arguing successfully that physicians should be able to prescribe contraception to preserve the health of their patients, even though contraception for birth control was illegal.[7]

McCormick's story is equally compelling. Born in 1875 into a wealthy family, she majored in biology and was one of the first two women to graduate from the Massachusetts Institute of Technology, establishing among other things the right of women students to take off their big flowery hats in chemistry lab. She graduated in 1902 and married Stanley McCormick, son of the inventor of the mechanical reaper and even richer than she was. But tragedy struck on their honeymoon when Stanley began to experience mental problems. He was subsequently diagnosed with schizophrenia and spent the rest of his life on the beautiful California estate Katharine built for him, cared for by male nurses and his personal physicians and soothed by musicians, gardens, and various ineffective treatments. Katharine visited the estate frequently, although she did not actually see him face to face for some seventeen years, and wrote loving letters detailing her activities in support of women's rights, a cause he also supported.[8] She battled his family over his custody because she felt that his exceptionally dif-

ABOVE LEFT Katharine Dexter McCormick, c. 1920
ABOVE RIGHT A silenced Margaret Sanger appeared onstage at Ford Hall Forum in Boston and handed her prepared speech to Harvard professor Arthur M. Schlesinger to read, April 16, 1929

ficult mother was at least partly responsible for his condition: his older sister was permanently institutionalized at nineteen, and two of his other siblings had serious mental problems as well.[9]

Because McCormick traveled several times a year to Europe, where she stayed at her family's chateau in Switzerland, she and several other wealthy supporters became an important source of birth-control devices—diaphragms—for Sanger's clinics. McCormick purchased trunk loads of fashionable clothes each season and hired seamstresses to fill their hems and hidden pockets with hundreds of diaphragms, all of which passed smoothly through customs since no official would suspect a woman of her standing of smuggling.[10] McCormick was particularly drawn to Sanger's program because they agreed that women would never be free until they could control their own fertility. Both McCormick and Sanger saw the American Birth Control League, founded by Sanger and Frederick Blossom in 1915 and later renamed Planned Parenthood, as the best way to accomplish this control.

Although McCormick supported Sanger's pursuits financially, she had always used her own money, never her husband's, which was spent on his care and on research into schizophrenia. But after Stanley McCormick died in 1947, Katharine became even wealthier. In 1948, when Pincus began discussions with Margaret Sanger about the possible development of a contraceptive pill for women, McCormick was determined to fund the development of the Pill. The project would eventually cost her some $3 million (about $28 million in 2007 dollars).[11] G. D. Searle also provided research funding to the Worcester Foundation as did Planned Parenthood, but most of the financial support for the Pill, especially after human trials began, came from McCormick.

The testing of the Pill is itself a controversial issue, since it was initially tested on psychotic women and men in a Massachusetts

Planned Parenthood was founded to establish clinics and to lobby against legislation prohibiting education about, or distribution of, contraceptives.

mental institution and on nurses in a nearby hospital. (The men were tested to see whether the female hormone calmed them down, but no evidence of such an effect emerged).[12] The Pill was also tested in Puerto Rico in 1956 and then in Haiti, Japan, Mexico, and England; the Puerto Rican tests led to U.S. Food and Drug Administration (FDA) approval of Enovid, containing norethynodrel, the synthetic progestin developed at Searle in 1957. Although both Enovid and Norlutin (developed simultaneously at Syntex) were initially approved for the treatment of such menstrual disorders as endometriosis, their contraceptive effects were reported by the pharmaceutical company to physicians, some of whom apparently prescribed them off label to their patients—half a million of whom were taking the Pill within two years of its 1960 release.[13]

Before 1962 the FDA was expected to examine new drugs for purity only, not for safety or efficacy. Still, Pasquale DeFelice, the young Roman Catholic physician to whom the agency outsourced the decision whether or not to approve the Pill, hesitated. On hearing that DeFelice had reservations, Rock flew to Washington to meet with him. When the doctor raised the issue of Catholic opposition to the Pill, Rock floored him with the challenge not to "sell short his church." In 1960 the FDA approved the Pill for contraceptive use, and the market exploded. By 1965, six-and-a-half million married women were taking the Pill.[14]

Planned Parenthood wanted the Pill for American women but also hoped to make it available around the world, especially to Third World nations whose growing populations threatened to destabilize their governments. The Pill was used in India and in some African nations, but it was considered both too expensive and too complicated for illiterate women, who were expected to begin taking it on the fifth day after their periods began and continue to take it for exactly twenty-two more days.[15] Although pharmaceutical companies sold the Pill to

Throughout history women have practiced many dangerous and ineffective methods of oral birth control, including the consuming of dried beaver testicles, mercury, and gunpowder.

·

Chrissy Conant
Chrissy Caviar,® 2001–2002
Installation view, Tang Museum

The 1969 book
*The Doctors' Case
against the Pill*
by Barbara Seaman
uses personal case
histories and medical
research to expose
the dangerous
side effects of early,
high-dose oral
contraceptives and
the failure of doctors,
pharmaceutical
companies, and the
U.S. Food and Drug
Administration
to inform women
about them.

Planned Parenthood International at a substantial discount, intrauterine devices and sterilization were much more important for international population control. Planned Parenthood in the United States opened clinics in inner cities so that the poor could better control family size, something that African American activists, including the Black Panthers, argued was a form of genocide and merely a racist effort to reduce the African American population.[16]

The story moves now in several different directions. What were the consequences of the Pill? Was the Pill safe? Who opposed the Pill? One obvious consequence of the Pill's introduction was a decline in the U.S. birth rate, apparent by 1957 (although demographers usually describe the baby boom as stretching from 1946 to 1964, when the number of births, as opposed to the birth rate, began to decline).[17] Sanger and McCormick lived long enough to see the reality of female-controlled contraception and the beginning of female empowerment, which made possible better educational and career opportunities and progress toward equity in employment.[18] A less positive consequence was to the health of American women: the initial versions of the Pill contained more than six times as much progestin as today's version. As early as 1962 some Pill users were exhibiting serious health problems, especially blood clots and strokes.[19] Those health problems led to a 1970 Senate hearing investigating the hazards of the Pill, a hearing disrupted by angry young women who believed that pharmaceutical companies had used them as guinea pigs for an unsafe drug. This hearing in turn led to the birth of the women's health movement.[20] A third consequence, according to some historians, was the sexual revolution.

The term *sexual revolution* refers to the loosening of sexual mores, the decline of the double standard in sexual attitudes toward men and women, and the increased visibility of sexuality, which apparently began in the 1960s. Pornography, swinging, premarital sex, and gay and lesbian civil-rights activity either threatened the end of civilization or were marks of human progress, depending on one's point of view.[21] And the Pill was very much a part of this popular culture.

But did the Pill cause the revolution? And was it a revolution anyway? We do know that premarital sex in America was much more common in the eighteenth century than in the nineteenth, part of a pattern of alternate peaks and troughs that was impressively documented in 1971 in an important study by Daniel Scott Smith and Michael Hindus.[22]

We also have some clues about changing patterns of premarital sex in the twentieth century: the Kinsey reports at least suggest that some women who reached maturity in the 1910s and 1920s were far more likely to experience sex before marriage, usually with their future husbands, than were women who reached maturity before 1900. But those studies focused on middle-class, college-educated women, and there is considerable evidence that working-class women began to increase their level of premarital sexual activity well before 1900.[23] Combining the Kinsey studies with smaller, earlier studies suggests that by the 1930s almost 90 percent of American women engaged in premarital petting (or foreplay) and perhaps half of them experienced premarital sex, again mostly with the men they would eventually marry. But the problem with these studies, all of middle-class, educated white women, is that they do not fully reflect the diverse experiences of American women, and as a recent *New York*

"Kay, you didn't tell me you gave up the pill, too!"

Cartoon from *Time*, March 9, 1970

Times story suggests, people responding to sex surveys do not always tell the truth. (One recent survey, administered to the same teenagers two years in a row, found that 10 percent of them seemed to have regained their virginity over time.)[24]

Another way to explore changing patterns of premarital sex is to look at illegitimacy statistics, which Hindus and Scott Smith found roughly parallel to the rates of bridal pregnancy in the eighteenth and nineteenth centuries.[25] (They also found an interesting pattern of class difference: premarital pregnancy had an inverse relationship to class status in the eighteenth century, with lower-class women having higher rates, although by 1960 this relationship had apparently disappeared.)[26] Although such records are not always totally accurate, as some mid-twentieth-century families were able to hide illegitimate births,[27] they do suggest quite interesting changes in the recent past. The 1982 National Survey of Family Growth included useful information about unmarried women and the Pill: between 1965 and 1969 only 5.7 percent of unmarried women used the Pill the first time they had sex, and the rate of abortion stood at 6.7 per 1,000 unmarried women.[28] George A. Akerlof, an economist, suggests that the greatest increase in Pill use by unmarried women seems to have occurred in 1969 and 1970.[29] From 1970 to 1974 the percentage of unmarried women who used the Pill the first time they had intercourse was 15.2 percent, while the rate of abortion had risen to 35.3 per 1,000. The years 1975 to 1979, after the *Roe v. Wade* decision that made abortions legal had gone into effect, saw Pill use at first intercourse for unmarried women drop to 13.4 percent and abortion numbers continuing to rise, reaching 50 for every 1,000 pregnant, unmarried women.[30] So more unmarried women were using the Pill, and more unmarried women were having abortions. But the vast majority of unmarried women who had sex did not use contraception at the outset, nor did they have abor-

In 1965 the U.S. Supreme Court ruled in *Griswold v. Connecticut* that the Connecticut law prohibiting the distribution of contraceptive devices to married couples was unconstitutional. The decision was later extended to apply to single women as well.

tions. Based on such statistics, equating the existence of the Pill with an increased likelihood for women to have premarital sex, especially since legal abortion apparently existed as a backup, seems too simple an explanation once we know that illegitimacy continued to rise, shotgun marriages began to decline, and many single young women had sex without using the Pill.

To further suggest that the Pill caused a sexual revolution in the 1960s is hard to prove, in part because all the demographic studies of Pill use in the decade were studies of married women. So we are left with retrospective and anecdotal information and a few early-1970s studies that are suggestive but not definitive. In 1971 only 28 percent of four thousand never-married women aged fifteen to nineteen reported having had sex; 20 percent of those having sex reported using the Pill, 27 percent condoms, and 24 percent the withdrawal method.[31]

In the 1960s unmarried women who wanted the Pill often had trouble getting it. Unlike earlier methods of contraception that could be purchased in drugstores or from mail-order sources or even made at home (e.g., douches and suppositories), the Pill required a doctor's prescription. Many doctors refused to prescribe the Pill to single women, especially if they were young, and many Planned Parenthood clinics, particularly those outside of urban areas, did not offer their services to unmarried women.

Colleges and universities in the 1960s struggled with policies regarding provision of the Pill, especially to undergraduates. In 1970 Planned Parenthood of Tompkins County opened a clinic on the Cornell University campus, an interesting policy evolution for an institution that only eight years earlier had suspended a male graduate student for cohabiting with a woman. A few years earlier a Harvard psychiatrist had warned colleges that allowing women to visit men in their dormitories was "aiding and abetting illegal and immoral activity,"

Scientists originally created two types of progestin, Norlutin and Enovid, with their only difference being the position of one double bond.

Melissa Gwyn
Progesterone, 1992
Ink, pencil, watercolor, gouache, and pastel
on paper
28 x 22 inches

blaming the decline of morals on women "seeking equality with men in all areas." In Lawrence, Kansas, the university health clinic would not prescribe the Pill to unmarried women, and the local Planned Parenthood clinic, open only two days a month, followed suit. But as early as 1965 one Lawrence doctor running his own clinic handed out the Pill to any woman who asked, often without even bothering to examine her.[32]

In 1970, at Skidmore College a student feminist group, Bread and Roses, installed a contraceptive exhibit in the library display cases complete with packages of Pills, diaphragms, containers of foam, and condoms. But despite the Bread and Roses exhibit, in the late 1960s and early 1970s the college physician would generally give the Pill only to students who were over twenty-one or about to be married, although the infirmary rules did indicate that contraceptives could be prescribed "at the doctor's discretion."[33]

Even today ambivalent attitudes toward the Pill remain. According to a recent article by Russell Shorto in the *New York Times Magazine*, the Pill is now being attacked as an abortifacient.[34] For the past fifty years the Pill has been generally understood to work by suppressing ovulation. It is modeled after the natural pattern in which the *corpus luteum*, developed in the ovarian follicle that produces the egg, secretes progesterone that suppresses further ovulation. But progestin (which mimics progesterone) also does two other things: it thickens cervical mucus, which then becomes a stronger barrier to sperm, and it apparently thins the endometrium, the blood-rich uterine lining that would otherwise be available for the implantation of the fertilized egg.

Pincus and Rock referred to the Pill as providing contraception primarily by suppressing ovulation and secondarily by thickening cervical mucus and thinning the endometrium. Even though half of all fertilized eggs do not implant and only half of the remainder actually

Before the 1930s physicians had so little understanding of the menstrual cycle that they recommended women have sex exactly between menstrual cycles to prevent pregnancy. This is now known to be the time when women are most fertile.

grow, antiabortion activists are now attacking the Pill—not just the abortion Pill (RU486), or the morning-after pill (which is simply a large dose of progestin), but the birth-control pill itself. They view the Pill as an abortifacient since it might impede the implantation of a fertilized egg, an event defined by such activists as ending a human life.[35]

Women who believe that the medical advances of the past half-century have given them control over their own fertility might indeed feel a chill down their spines at this latest development. Margaret Sanger, Katharine McCormick, Gregory Pincus, and John Rock, as well as the others who worked to develop the Pill, might well be rolling over in their graves.

1. Lara V. Marks, *Sexual Chemistry: A History of the Contraceptive Pill* (New Haven, CT, & London: Yale University Press, 2001), 44.

2. Bernard Asbell, *The Pill: A Biography of the Drug That Changed the World* (New York: Random House, 1995), 110.

3. Loretta McLaughlin, *The Pill, John Rock, and the Church: The Biography of a Revolution* (Boston: Little, Brown, 1982), 113.

4. Ibid., 109–110.

5. Elizabeth Siegel Watkins, *On the Pill: A Social History of Oral Contraceptives, 1950–1970* (Baltimore & London: Johns Hopkins University Press, 1998), 29.

6. The Massachusetts statute had originally been passed in 1879, in the wake of federal legislation (the "Comstock" law) that outlawed sending obscenity through the mails and designated contraceptive information and contraceptive devices as legally "obscene." When a similar statute was declared an unconstitutional imposition on the privacy of married couples by the U.S. Supreme Court in *Griswold v. Connecticut* (381 U.S. 479) in 1965, the Massachusetts legislature had amended its statute to allow doctors to prescribe contraception and pharmacists to fill such prescriptions but only for married couples. Only after the Supreme Court extended the right of privacy to singles in *Baird v. Eisenstadt* (405 U.S. 438 [1972]) was the Massachusetts statute eliminated. Wisconsin was the last state in the nation to eliminate laws against contraception, which repealed its Comstock-era statute banning the provision of contraception to the unmarried in 1975, three years after the Baird decision. From Planned Parenthood of Wisconsin, Inc., "Our History," www.ppwi.org/aboutus/history/ourhistory.aspx (accessed June 6, 2007).

7. Linda Gordon, *Woman's Body, Woman's Right: Birth Control in America* (New York: Penguin, 1990), 253–255.

8. Some historians have attributed Katharine's interest in contraception to her concern that she not bear children who might have inherited their father's illness and refer to Stanley's sexual demands, but the fullest and most recent biography (Armond

Fields, *Katharine Dexter McCormick: Pioneer for Women's Rights* [Westport, CT: Praeger, 2003]) documents Stanley's impotence (pp. 65 and 156), his terror of women, and the fact that he was not allowed to see his wife for the seventeen years from 1908 to 1925, although the two carried on regular correspondence (193 and 208).

9. Ibid., 39–41, 69.

10. Ibid., 181–182.

11. Katharine McCormick may have hesitated to spend Stanley's income on projects she supported for feminist reasons because of the long battle she had waged with his relatives over custody of her husband and control of his estate. Once he had died and she inherited his millions, she was free to use the funds as she saw fit.

12. McLaughlin, *The Pill* (cit. note 3), 119–120.

13. N. Oudshoorn, *Beyond the Natural Body: An Archaeology of Sex Hormones* (London: Routledge, 1994), 133; quoted in Marks (cit. note 1), 111.

14. Watkins, *On the Pill* (cit. note 5), 34.

15. Ibid., 70–71.

16. Ibid., 56.

17. Watkins (cit. note 5), 63; Louis B. Russell, *The Baby Boom Generation and the Economy* (Washington, DC: Brookings Institution, 1982), 2–3.

18. Sanger died in 1966, McCormick in 1967.

19. Barbara Seaman, *The Doctor's Case against the Pill* (New York: Peter H. Wyden, 1969), 89, 107–108.

20. Watkins (cit. note 5), 107–119, *passim*.

21. Estelle Freedman and John D'Emilio, *Intimate Matters* (New York: Harper & Row, 1988), ch. 13, *passim*.

22. Michael Hindus and Daniel Scott Smith, "Premarital Pregnancy in America 1640–1971: An Overview and Interpretation," *Journal of Interdisciplinary History* 5:4, "The History of the Family, II" (Spring 1975), 537–570.

23. Freedman and D'Emilio, *Intimate Matters* (cit. note 21), 256–260, *passim*; Daniel Scott Smith, "The Dating of the American Sexual Revolution: Evidence and Interpretation," quoted in Michael Gordon, *The American Family: Past, Present, and Future* (New York: Random House, 1978), 190–191.

24. Eric Nagourney, "Patterns of Deceit Raise Concerns about Teenage Sex Surveys," *New York Times*, May 9, 2006, section F, p. 5.

25. Demographers define *bridal pregnancy* as the birth of a living child born less than seven months after a couple's wedding, that is, a child probably conceived sometime before the ceremony.

26. Hindus and Scott Smith (cit. note 22), 541–544.

27. This would often involve sending a pregnant daughter out of town to a maternity home where she would give birth and turn the infant over to an adoption agency. See Ann Fessler, *The Girls Who Went Away* (New York: Penguin, 2006).

28. The National Surveys of Family Growth are put out by the National Center for Health Statistics of the Department of Health and Human Services.

29. 1982 National Survey of Family Growth, quoted in George Akerlof et al., "An Analysis of Out-of-Wedlock Childbearing in the United States," *Quarterly Journal of Economics* 111:2 (May 1996), 288.

30. Ibid., 284.

31. John F. Kantner and Melvin Zelnick, "Sexual Experience of Young Unmarried Women in the United States," *Family Planning Perspectives* 4 (October 1972), 9, cited in Watkins, *On the Pill* (cit. note 5), 63.

32. Beth Bailey, "Prescribing the Pill: Politics, Culture, and the Sexual Revolution in America's Heartland," *Journal of Social History* 30:4 (Summer 1997), 827–857, *passim*.

33. Skidmore College, *Student Information Booklet*, 1974 (Skidmore College Archives).

34. Russell Shorto, "Contra-Contraception," *New York Times Magazine*, May 7, 2006, 48–55, 68, 83

35. Randy Alcorn, "Does the Birth Control Pill Cause Abortions?" Eternal Perspectives Ministries, www.epm.org /articles/bcp5400.html (accessed May 8, 2006).

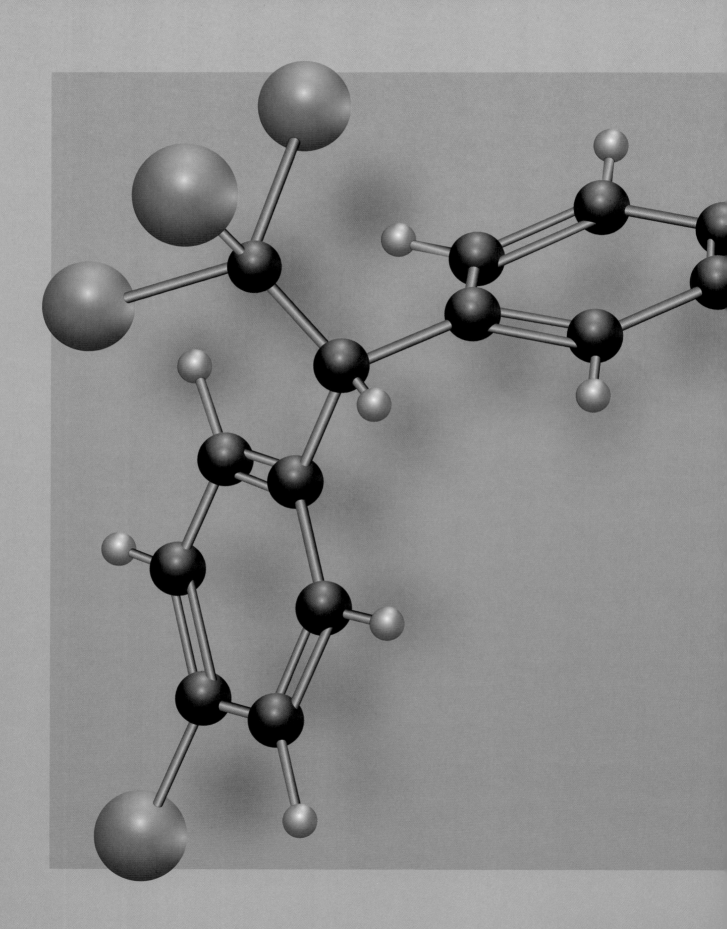

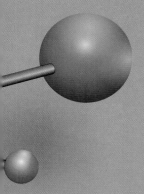

DDT

$$C_{14}H_9Cl_5$$

U.S. PATENT NUMBER: 2,654,790
CAS REGISTRY NUMBER: 50-29-3

DDT, or 2,2 bis (p-chlorophenyl) 1,1,1-trichloroethane, is one of this century's most controversial molecules. It is either loved for its insecticidal properties or hated for its environmental persistence and biomagnification properties. There are significant generational and geographic differences of opinion about DDT and its uses. If you were alive in the 1930s, 1940s, and 1950s, you probably are glad it was discovered, but if you were born after 1960, you are likely to have a negative opinion of DDT. If you live in North America or western Europe, you probably want very restricted use of DDT, but if you live in the tropical regions of Asia, South America, or Africa, you might wish it were more available to you.

DDT was first synthesized in 1874 in Strasbourg, Germany, by Othmar Zeidler as part of his doctoral dissertation, although he failed to recognize it as an insecticide.[1] In 1939 the Swiss chemist Paul Mueller, working for J. R. Geigy AG, recognized its insecticidal properties.[2] His discovery of DDT's effectiveness in controlling insect-borne diseases earned him the 1948 Nobel Prize in Physiology or Medicine.

by Robert J. Hargrove

Before the discovery of DDT and other chlorinated hydrocarbons the principal substances used for insect control were pyrethrins and arsenicals. During the early years of World War II, European nations and the United States had difficulty getting these substances because the German and Japanese navies controlled the seas, making the importation of all goods difficult. It was already well known that arsenicals were very toxic to humans, and DDT proved an effective alternative because it is so simple to make.[3]

Prior to World War II the outcome of most wars was determined by which side had the most survivors after malaria, yellow fever, typhus, or some other scourge had taken its victims. This problem had been true throughout the centuries: disease killed more troops than bullets in the American Revolutionary War and the Civil War. During World War II, for the first time in history the total deaths from disease were lower than the total deaths from wounds.

In 1939 the German armies were marching into Poland, and Switzerland was having trouble receiving chemical shipments from other countries. The Colorado potato beetle, which had been introduced into Europe with the U.S. Army's supplies during World War I, struck the Swiss potato crop. The Swiss government used DDT as a pesticide, and it passed its first major test by destroying the beetle and saving the crop. Further tests showed it was effective against many other insects, including lice.

Large outbreaks of insect-vectored diseases affected both military and civilian populations during World War II. During a typhus epidemic in Italy in the early 1940s, DDT was used to delouse soldiers and civilians, preventing the spread of the disease. In the Pacific and in Africa at this time malaria was

In 1954 there were two million cases of malaria in Greece; by 1972, after the widespread use of DDT there were only seven known cases.

Nr. 226180   Klasse **3c**

SCHWEIZERISCHE EIDGENOSSENSCHAFT

EIDG. AMT FÜR GEISTIGES EIGENTUM

PATENTSCHRIFT

Veröffentlicht am 16. Juni 1943

Gesuch eingereicht: 7. März 1940, 19 Uhr. — Patent eingetragen: 31. März 1943.

HAUPTPATENT

J. R. GEIGY A.-G., Basel (Schweiz).

Verfahren zur Schädlingsbekämpfung.

Für die Bekämpfung von Insekten aller Art, wie Fliegen, Stechmücken, Motten, Käfern, Blattläusen etc., sowie deren Entwicklungsstadien, werden hauptsächlich Petroleumlösungen von Pyrethrin oder Rotenon oder wäßrige Emulsionen dieser Verbindungen verwendet. Nikotin findet trotz seiner Giftigkeit für den Menschen im Pflanzenschutz Verwendung, scheidet aber für den Gebrauch in bewohnten Räumen aus.

Die ersten beiden Mittel haben den Nachteil, daß sie in Form von Petroleumlösungen trotz Beimischung von starken Parfümierungsmitteln unangenehm riechen. In wäßriger Emulsion dagegen sind sie nur wenig haltbar, ihre Wirksamkeit nimmt schon nach kurzer Zeit stark ab.

Alle Versuche, künstliche Stoffe aufzufinden, die sehr rasch und sicher wirken, dabei fast oder ganz geruchlos sind und keinerlei Reizwirkung auf den Menschen ausüben, hatten bis heute keinen wesentlichen Erfolg. So ist beispielsweise auch die Verwendung von halogenierten Nitrilen, insbesondere Tri-

chloracetonitril, auf unbewohnte Gebäude oder auf geschlossene Behälter beschränkt, weil diese halogenierten Verbindungen auch in geringster Verdünnung äußerst heftig auf die Augenschleimhäute einwirken.

Um so überraschender ist es, daß Verbindungen von der allgemeinen Formel

$$R_1 - CH - C \underset{X}{\overset{X}{\langle}} X$$
$$\quad\quad R_2$$

in welcher X Halogen, R_1 einen organischen Rest mit mindestens 3 Kohlenstoffatomen und R_2 einen organischen Rest mit mindestens 6 Kohlenstoffatomen bedeuten, neben einer sicheren tötlichen Wirkung auf Schädlinge, insbesondere Insekten, nur einen sehr schwachen und keineswegs unangenehmen Geruch aufweisen und auch in feinverteilter Form keinerlei Reizwirkung auf die Nasen-, Augen- oder Rachenschleimhaut ausüben.

Die nach vorliegendem Verfahren zur Anwendung gelangenden Verbindungen der an-epidemic but preventable. Large tracts of land were sprayed with DDT to kill mosquitoes, which transported the disease from infected humans to healthy ones. Brigadier General James S. Simmons said, "DDT is one of the most wonderful things, medically speaking, to come out of the war, but it must be used with intelligence and judgment."[4]

The American Association of Economic Entomologists issued a statement on December 15, 1944:

> We feel that never in the history of entomology has a chemical been discovered that offers such promise to mankind for relief from his insect problems as DDT. There are limitations and qualifications, however. Subject to these, this promise covers three chief fields: public health, household comfort, and agriculture.
>
> As public health we include control of the insects which carry diseases that have scourged humanity, such as malaria, typhus, and yellow fever. Household comfort is taken to cover such things as flies, bedbugs, and mosquitoes. Agriculture includes not only farms, gardens, and orchards, but forests, livestock, and poultry.[5]

DDT is certainly an effective insecticide. Insects absorb it through their exoskeleton by direct contact, so spraying surfaces where they rest, such as walls of homes, is an effective treatment. As a neurotoxin, the proposed sites of action on the synapse are reducing potassium

OPPOSITE Swiss patent for DDT granted to J. R. Geigy Company, March 31, 1943
ABOVE LEFT TO RIGHT Neocide DDT advertisement, France, c. 1940; L'Insectoline DDT advertisement, France, c. 1930

transport through pores; inactivating sodium channel closure; inhibiting sodium-potassium and calcium-magnesium ATPases; and calmodulin-calcium binding with release of neurotransmitter.

DDT soon suffered from its own success, however. People thought that if using a small amount worked well, then using more would work even better. In a matter of a few years DDT was widely available and was used to combat insects around the world. Unfortunately it killed the good insects as well as the harmful. It was being sprayed across forests and fields, in homes and offices, on children's heads to control head lice in schools, in kitchens to control flies. It indiscriminately killed beetles, moths, lice, butterflies, flies, and mosquitoes. By 1959 DDT manufacture rose to nearly 125 million pounds annually for disease control, agriculture, gardening, and animal husbandry.[6] It was used throughout the world but most heavily in the tropical regions.

With such widespread use many harmful insects quickly became resistant to DDT. As fewer insects died, the applicators just used greater amounts of DDT. Yet little was known about the pesticide's persistence and biomagnification properties in the environment until Rachel Carson, a biologist working for the U.S. Bureau of Fisheries (which later became the U.S. Fish and Wildlife Service) published her third book, *Silent Spring*, in 1962.

Silent Spring affected the way the entire world viewed chemical pesticides, and its impact was immediate and dramatic. The book's major theme is the interdependence of life on earth, and Carson argued convincingly that polluting the ecosphere with pesticides or other toxic compounds results in unintended and unforeseen consequences. The book correctly blamed reduction in bird, fish, and certain mammal populations on indiscriminate DDT use.

An unforeseen decline in the number of bald eagles, peregrine falcons, ospreys, brown pelicans, and other birds of prey motivated

DDT is estimated to have saved the lives of approximately fifty million people.

·

Melissa Gwyn
DDT, 2007
Oil on wood panel
36 x 24 x 2 inches

Carson to write *Silent Spring*. DDT is very slowly metabolized by the loss of hydrogen chloride to dichlorodiphenyldichloroethylene (DDE), which is stored in the fatty portions of an organism.[7] This substance is said to have a biological half-life of nearly forty years. As insects are killed by DDT and earthworms absorb it through the soil, small birds eat the insects and worms, absorbing the DDT and its metabolite DDE. Then as small birds become prey for large ones, the DDT becomes further concentrated the higher it travels up the food chain. For example, if seawater contains 0.003 parts per billion (or 0.000003 parts per million) of DDT, then the zooplankton may have 0.04 parts per million; the small fish that eat the zooplankton may have 0.5 parts per million, larger fish that eat small fish may have 2 parts per million, and a fish-eating bird like an osprey can have as much as 25 parts per million in its tissues.[8] At these high levels DDT has been shown to affect the calcium metabolism of birds, weakening their eggshells, which causes most eggs to break in the nest during incubation and ultimately results in a decline in the population.

Rachel Carson's 1962 book *Silent Spring* helped launch the environmental movement.

Silent Spring was widely read and stimulated both the public and politicians to become involved in this issue. A group of lawyers started an effort to ban DDT, and then in 1967 they incorporated as the Environmental Defense Fund. At that time many large varieties of persistent pesticides were being used and were thus contaminating the environment. Concern about industrialized societies' level of pollution escalated all over North America and Western Europe. In 1970 President Richard M. Nixon formed the U.S. Environmental Protection Agency (EPA) whose mission is "to protect human health and the envi-

Silent Spring, Rachel Carson, first edition
Houghton Mifflin, 1962; Frank Graham, Jr., *Since
Silent Spring*, Houghton Mifflin, 1970

ronment." Regulations and laws were enacted to reduce the level of exposure to harmful substances. However, DDT remained controversial because there was little evidence that it was harmful to humans. Although by the 1970s its effect on wildlife was well established, many argued that its benefits to humans by preventing disease far outweighed the adverse effects on wildlife.

After three years of government investigation into the uses of DDT, William D. Ruckelshaus, administrator of the EPA, ultimately concluded that "the continued use of DDT posed unacceptable risks to the environment and potential harm to human health." On December 31, 1972, the EPA issued a statement banning almost all domestic use of DDT. After the ban in the United States, DDT continued to be manufactured and used in many countries, particularly those in Asia, South America, and Africa where insect problems remained paramount.

Because DDT and other halogenated substances (hydrocarbons that contain multiple chlorine or bromine atoms) are so persistent, humans and wildlife continued to be exposed. Every time soil contaminated with DDT was tilled, DDT would escape into the air and be blown downwind. It was also found that even though most of the DDT was used in tropical regions, its concentration in the environment was widespread, even reaching the polar regions. In the tropical regions DDT would evaporate and be transported to cooler regions where it would then condense, exposing fish and birds to it. The larger polar animals, such as seals, walruses, penguins, and polar bears, which ingest fish and birds, subsequently became exposed to high levels of DDT and its metabolite DDE.

By 2000 the United Nations Environment Programme had completed negotiations on a treaty to ban the production and use of what has become known as the "dirty dozen" persistent organic pollutants, of which DDT is the most well known.[9] This treaty, known as the

DDT kills mosquitoes by acting as a nerve poison; it is highly fat-soluble and is therefore found in such fatty foods as meat and dairy products.

DDT
Installation view, Tang Museum

Robert Dawson
200 tons of DDT buried underwater, Santa Monica Bay, CA, 1989
Gelatin silver print
16 x 20 inches

Stockholm Convention on Persistent Organic Pollutants, has been signed by 152 countries and ratified by 136. The United States has yet to ratify this convention.

Malaria continued to be one of the most deadly infectious diseases after World War II, and DDT was viewed as the primary agent for its eradication. But, as previously stated, the development of resistance in insects slowed this effort. Since the early 1970s the incidence of malaria has slowly and progressively increased. Reduced control measures owing to financial constraints between 1972 and 1976 led to a massive two- to threefold rise in malaria cases globally. Spraying insecticide never truly eradicated mosquitoes anywhere, and infections returned in much greater strength as control measures waned. Increased movement of people in and out of malarial regions owing to commerce and civil conflict exposed more people to the mosquitoes transmitting the disease.

In the 1980s scientists discovered that DDT and its residues at levels of parts per billion are endocrine disrupters. These substances, along with many other synthetic chemicals, act as environmental estrogens, meaning that exposure during embryonic development causes a fetus either to become female or to adopt feminine characteristics. Exposure to DDT and the incidence of breast cancer also seem to be linked. Epidemiological studies on this subject have been inconclusive; however, the 1996 book *Our Stolen Future* discusses the effects of DDT and other environmental estrogens on humans and wildlife.[10]

The lethal dose of DDT for humans is about 30 grams (one ounce).

Today malaria remains the world's most widespread infectious tropical disease despite all the measures used to control it. The World Health Organization (WHO) reports that more than five hundred million people suffer from acute malaria, resulting in more than one million deaths each year. At least 86 percent of these deaths are in sub-Saharan Africa. Globally an estimated three thousand children

In 2006 the bald eagle, one of the birds of prey affected by large amounts of DDT in the environment, was taken off the endangered species list.

and infants die from malaria every day, and ten thousand pregnant women die from malaria in Africa every year, with almost 60 percent of the cases occurring among the poorest 20 percent of the world's population.[11] Because of this situation, in September 2006, thirty years after WHO began phasing out the use of DDT and other insecticides to control malaria, the organization issued a statement once again recommending the use of indoor residual spraying of DDT in areas of high malaria transmission. "Indoor residual spraying [IRS] is useful to quickly reduce the number of infections caused by malaria-carrying mosquitoes," said Anarfi Asamoa-Baah, the WHO assistant director-general for HIV/AIDS. "IRS has proven to be just as cost effective as other malaria prevention measures, and DDT presents no health risk when used properly."[11] IRS involves applying long-acting insecticides on the walls and roofs of houses and animal shelters in order to kill malaria-carrying mosquitoes that land on these surfaces.

The EPA has also reapproved indoor residual spraying despite its effects on birds of prey and on humans. Through careful study and discriminating use DDT is still an effective, inexpensive insecticide for the control of mosquitoes. Yet if this celebrated and controversial molecule is to remain in the arsenal of weapons used to combat malaria and other diseases, we must find ways to harness DDT's beneficial effects, while mitigating its deleterious ones.

1. Othmar Zeidler, "Verbindungen von Chloral mit Brom- und Chlorbenzol," *Chemische Berichte* 7 (1874), 1,180–1,181.

2. Swiss Patent No. 226180, J. R. Geigy, Inc., "Verfarhen zur Schaedlings-bekaempfung," submitted March 7, 1940; issued March 31, 1943.

3. Eugene L. Bailes published this synthesis: Mix 16.64 g chloral hydrate, 22.5 g chlorobenzene, and 200 mL concentrated sulfuric acid in an oil bath with mechanical stirrer heating for 8 to 10 hours not to exceed 105°C. Wash multiple times in water to remove excess sulfuric acid and recrystallize from alcohol. See "Method for the Preparation of 1,1,1-Trichloro-2,2-Bis(p-Chlorophenyl) Ethane Commonly Known as DDT," E. L. Bailes, *Journal of Chemical Education* 22 (1945), 122.

4. James C. Leary, William I. Fishbein, and Lawrence C. Salter, *DDT and the Insect Problem* (New York: McGraw-Hill, 1946), v.

5. Ibid.

6. Thomas R. Dunlap, *DDT: Scientists, Citizens, and Public Policy* (Princeton, NJ: Princeton University Press, 1981), 254.

7. Curtis D. Klaassen, Mary O. Amdur, and John Doull, "DDE: dichlorodiphenyl-dichloroethylene," in *Casarett and Doull's Toxicology: The Basic Science of Poisons*, 5th edition (New York: McGraw-Hill, 1996), 127.

8. The persistent organic pollutants are
• Aldrin: insecticide used against soil pests (primarily termites) on corn, cotton, and potatoes.
• Chlordane: insecticide now used primarily for termite control.
• DDT: insecticide now used mainly against mosquitoes for malaria control.
• Dieldrin: insecticide used on fruit, soil, and seed crops, including corn, cotton, and potatoes.
• Endrin: rodenticide and insecticide used on cotton, rice, and corn.
• Heptachlor: insecticide used against soil insects, especially termites; also used against fire ants and mosquitoes.
• Hexachlorobenzene: fungicide; also a by-product of pesticide manufacturing and a contaminant of other pesticide products.
• Mirex: insecticide used on ants and termites; one of the most stable and persistent pesticides; also a fire retardant.
• Toxaphene: insecticide used especially against ticks and mites; a mixture of up to 670 chemicals.
• Polychlorinated biphenyls (PCBs): used primarily in capacitors and transformers and in hydraulic and heat transfer systems; also used in weatherproofing, carbonless copy paper, paint, adhesives, and plasticizers in synthetic resins.
• Dioxins: by-products of combustion (especially of plastics) and of chlorine product manufacturing and chlorine bleaching of paper.
• Furans: by-products, especially of PCB manufacturing, often with dioxins.

9. Roll Back Malaria, "2001–2010 United Nations Decade to Roll Back Malaria: Malaria in Africa," www.rbm.who.int/cmc_upload/0/000/015/370/RBMInfosheet_3.htm (accessed June 18, 2007).

10. Theo Colburn, Dianne Dumanoski, and John Peterson Myers, *Our Stolen Future* (New York: Dutton, 1996).

11. World Health Organization, "WHO Gives Indoor Use of DDT a Clean Bill of Health for Controlling Malaria," www.who.int/mediacentre/news/releases/2006/pr50/en/.

1980

PROZAC

$C_{17}H_{18}ONF_3$

U.S. PATENT NUMBER: 5,104,899
CAS REGISTRY NUMBER: 59333-67-4

The forced swim test, one of the most reliable preclinical screening tests used by pharmaceutical companies for prospective antidepressants, was first developed by Roger Porsolt in the mid-1970s. In this test a mouse is placed for several minutes in a cylinder of lukewarm water from which there is no escape. At first the mouse struggles and swims, but eventually it displays an immobility posture characterized by just enough movement to keep afloat. Researchers then rescue the animal and retest it twenty-four hours later. During the second session the mouse gives up much sooner, seeming to recall the hopeless nature of the task. Researchers call this phenomenon "behavioral despair" or "learned helplessness" and use it as an animal model for human depression. But if a researcher gives Prozac to the mouse before retesting, it struggles with renewed vigor, swimming for a significantly longer period before giving up. Virtually every antidepressant that enters human clinical trials has passed the forced swim test. How should the behavioral change that occurs with Prozac be interpreted? Does the drug strengthen resolve in the face of an unalterable reality?

by Hassan H. López

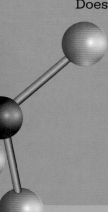

The fact remains that there is no escape from the cylinder, so the hope provided by Prozac is, from an existential standpoint, false. More important, what does this animal model say about Prozac's effect on people? Is the proper analogy for human life not a rat in a maze but rather a mouse in a water-filled cylinder?

In the early 1990s great changes occurred in the field of neuroscience. The human brain had long seemed an impenetrable mystery; our thought processes and behavior were considered too complex for any mechanistic framework. Yet new methodological breakthroughs provided scientists with the tools necessary to tie activity of the nervous system to specific psychological phenomena. Functional magnetic resonance imaging allowed for detailed, nonintrusive mapping of a working human brain. Genetic models linked the activity of individual genes with precise components of our behavioral repertoire. But perhaps the fastest advances were made in the field of psychopharmacology. By the late twentieth century hundreds of drugs had become available for the treatment of severe mental illnesses, as well as for more moderate clinical syndromes like anxiety and insomnia. For many Americans, manipulating brain chemistry had become part of their daily routine.

Around this time Prozac began to weave its way into the fabric of our society, with a cultural impact not witnessed since the introduction of Valium and the benzodiazepines forty years earlier. Perhaps more than any other drug Prozac destigmatized mental illness and pharmacotherapy. It was safe and gentle and promised relief for a wide spectrum of disorders. Yet as the twentieth century closed and more research accumulated, the hope and enthusiasm generated by Prozac slowly dissipated. It was not the ultimate solution to all our ills. Nevertheless, Prozac helped bring neuroscience and psychopharmacology into everyday conversation: it challenged our epistemological

Prozac is the most widely prescribed antidepressant drug in history, with 60 million prescriptions written since it was approved by the FDA in 1988.

assumptions and forced us to consider how playing with the human mind might affect the evolution of our civilization.

Prozac's sweeping impact on the treatment of depression should not be understated. Introduced by Eli Lilly in early 1988, it was the first in a new generation of antidepressant medications. Within two years Prozac was the most prescribed antidepressant in the world. Many people falsely attribute this success to an enhanced, nearly magical therapeutic effect experienced when using Prozac in comparison to the results achieved from previously available treatments. However, hundreds of clinical trials conducted over the last two decades have repeatedly demonstrated that Prozac shows no greater clinical efficacy in the treatment of depressive disorders than any other medication. Rather, the primary advantage of Prozac lies in its safety and relatively benevolent side-effect profile.

EDGAR ALLAN PROZAC

Prior to the introduction of Prozac, depression had been treated predominantly with either a monoamine oxidase inhibitor (MAOI) or a tricyclic antidepressant (TCA). Both MAOIs and TCAs are just as effective as Prozac in reducing depressive symptoms and sometimes show greater success in treatment-resistant patients. The MAOIs, however, induce a number of unpleasant symptoms, such as insomnia and severe headaches, and interact dangerously with other pharmaceuticals. Moreover, if a patient on a classic MAOI consumes foods high in tyramine (found in such items as aged cheese, smoked meat, and beer), a drastic and potentially fatal increase in blood pressure can occur. TCAs are even worse, with side effects ranging from the irritating to the harmful, including dry mouth, blurred vision, constipation,

Cartoon from *The New Yorker*, November 6, 2006

confusion, sedation, weight gain, dizziness, and hypotension. TCAs also carry a dangerous risk for overdose, severely damaging cardiac tissue at higher doses.

In comparison Prozac was a wonder drug. There was effectively no threat of overdose, and the typical side effects were mild in comparison: some nausea, periodic sleep disturbances, changes in appetite, and sexual dysfunction. For many patients these symptoms dissipate after a few weeks as the body develops a slight tolerance to the drug. The medical community saw these advantages as nothing less than revolutionary. Within a few years several Prozac mimics, including Zoloft, Luvox, Paxil, and Celexa, were introduced by competing pharmaceutical companies. Doctors increasingly prescribed Prozac—off label at first—for such nondepressive disorders as social anxiety, obsessive-compulsive disorder, post-traumatic stress disorder, panic disorder, bulimia nervosa, attention-deficit hyperactivity disorder (ADHD), fibromyalgia, and premature ejaculation. The mystique of Prozac grew, reaching its peak with Peter Kramer's book *Listening to Prozac* (Viking, 1993), in which he documented numerous case studies evincing radical, positive changes in personality among those taking the drug. The phrase "better than well" was introduced to our culture, and the inevitable comparisons to Aldous Huxley's pacifying, mind-controlling drug, soma (as described in his novel *Brave New World*), began to appear. Suddenly our existential sense of self was at stake, and perhaps for the first time the public became aware that advances in clinical pharmacology were going to have an effect on mental health.

Prozac's impact on society must ultimately be reduced to its effect on individual

Prozac was introduced in the United States in January 1988. By 1993, eight million people had taken Prozac; over four million of them were American.

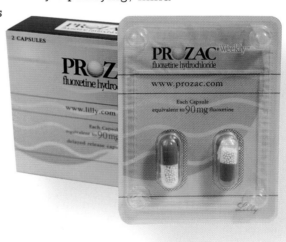

Melissa Gwyn
Prozac, 2007
Oil on wood panel
16 x 11 inches

ABOVE Prozac pills, 2007

human experience. The modern field of neuroscience provides us with a theoretical framework for understanding how drugs have the ability to affect our most intimate psychological experiences through manipulation of the brain's activity. Animal nervous systems have evolved in order to process information, allowing us to take in sensory input from the external environment, make decisions based on that information, and execute behaviors to accomplish specific goals. Any psychological process (such as deciding where to find a good burrito or pondering the meaning of a complex sentence) can be represented by the activity of a discrete network of neurons. *Mind* refers to the cognitive processes and phenomenological experiences that typify human psychology, while *brain* refers to the biological substrates of these mental operations. Any stimulus that affects brain activity, like a psychotropic drug, therefore has the potential to alter our psychological reality.

Most psychiatric medications (and recreational drugs) specifically affect how neurons communicate with each other. Individual neurons within a network are separated by small gaps, or synapses. When one neuron wishes to communicate with a neighboring neuron, it releases thousands of neurotransmitter molecules into the synapse. These molecules diffuse through extracellular space until they bind to specialized receptors on the receiving neuron. The human brain uses several dozen different neurotransmitters, each associated with a variety of psychological experiences, such as hunger, vigilance, sexual desire, and attachment. Prozac targets a single neurotransmitter system: serotonin. This specificity is what makes Prozac fundamentally different from its

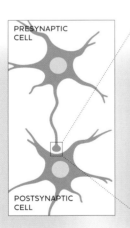

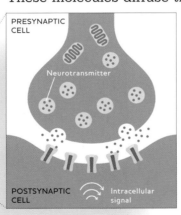

Schematic of two neurons communicating via a synapse

predecessors. The MAOIs and TCAs affect a bevy of chemical messengers, including dopamine, norepinephrine, epinephrine, serotonin, acetylcholine, and histamine. It is therefore not surprising that both the MAOIs and TCAs induce such a wide variety of physiological and psychological side effects in addition to their antidepressant activity.

Neurons typically release their neurotransmitters from a particular location in the cell called the axon terminal. Embedded in the cellular membrane of the axon terminal are specialized transporter proteins that allow the neuron to reabsorb neurotransmitter molecules located in the synapse. This system provides the neuron with a way to stop the signal it has been sending, as well as to recycle and reuse its messengers. The process is known as reuptake. Prozac functions by binding to the serotonin transporter and preventing it from carrying serotonin molecules across the cell membrane so that whatever serotonin is released into the synapse stays there for a significantly longer time. For this reason Prozac and similar compounds are called selective serotonin-reuptake inhibitors, or SSRIs.

But why does increasing serotonergic activity affect mood and lift depression? The simplest theory is that clinical depression is caused by a deficiency in serotonin and this deficiency is alleviated by SSRI administration. This serotonin hypothesis of depression has influenced clinicians and researchers for the past several decades and is supported by several empirical findings. Giving a drug that depletes serotonin reserves to patients in remission from depression induces depressive episodes. Recent neuroimaging studies have demonstrated that depressed patients show reduced activity within a part of the brain known as the prefrontal cortex, a primary target of serotonergic neurons. Some studies have found reduced levels of a serotonin metabolite, 5-HIAA, in the cerebrospinal fluid and blood plasma of depressed patients compared with nondepressed control patients.

Why higher serotonin levels are effective in reducing the severity of depression is still unknown.

LOST IN A MAZE
OF BEWILDERING
THOUGHTS

PROZAC®
fluoxetine hydrochloride

Weekly™

prozac

Prozac
Installation view, Tang Museum

Man, by his very nature, must live with anxiety

Chance made us sisters. Prozac made us friends.

EDGAR ALLAN PROZAC

PROZAC
SOMETIMES YOU FEEL LIKE A NUT
SOMETIMES YOU DON'T!

Birth
Ritalin
Prozac
Viagra
Death

I TOOK MY PROZAC TODAY

prozac makes it better

I TAKE MY VIAGRA WITH PROZAC... IF IT DOESN'T WORK... I DON'T CARE

got prozac?
truth-now.com WAV

Other studies have found no significant difference or find a difference only when looking at profoundly suicidal patients.

The serotonin hypothesis, however, is problematic. Since the introduction of SSRIs clinicians have noted a significant time lag, often weeks, between the first administration of the drug and the lifting of depressive symptoms. In comparison, a patient taking Valium for the treatment of anxiety or Ritalin for ADHD experiences the effects within an hour. SSRIs operate by significantly increasing serotonin activity in the brain within hours of administration. But if depression is based on a mere deficiency in serotonin, improvement should occur as soon as that deficiency is reversed. Recent research implicates the neurotransmitter, norepinephrine, in the pathophysiology of depression to just as great an extent as serotonin. Several successful antidepressants on the market today (e.g., Vestra and Remeron) are designed to target norepinephrine activity and do not significantly affect serotonin. Finally, Prozac does not work for all patients with depression: the maximum success rate for any antidepressant is approximately 65 percent. This rate reflects the fact that depression is a heterogeneous disorder in which there are subtle variations in syndromes among different patients, mediated by subtle differences in underlying neurochemical pathology (e.g., a serotonin deficiency in one patient and a norepinephrine deficiency in another).

Beneath these legitimate questions and concerns lies a deeper issue. If we accept the doctrine of modern neuroscience that all psychological phenomena are rooted in biological processes, then we must confront the possibility that Prozac alters our psychological reality in ways that are beyond our current understanding. Consider for a moment the very concept of depression: a negative mood characterized by sadness, loss of hope, reduced energy, guilt, and possibly thoughts of suicide. The emotional pain experienced seems to justify

Prozac is the first SSRI and is prescribed for treating depression, obsessive-compulsive disorder, bulimia nervosa, and panic disorder.

Thomas Asmuth
Fluoxetine (Prozac), 2006
Soft sculpture
60 x 40 x 20 inches

medication. Yet we prescribe medication without a concrete understanding of why the mind is even capable of such a mental state. Is the human mind unique in its ability to suffer profound depression? Do other animals experience a feeling that, if not identical, is at least qualitatively similar? What specific information-processing systems are affected by the depressed condition? Is phenomenological depression a result of an internal system malfunction or an external cause? The serotonin hypothesis and other purely neurological theories of depression cannot answer these questions because they are fundamentally atheoretical in terms of underlying etiology. Consider a patient who suffers a heart attack. We might rightfully argue that a buildup of arterial cholesterol is causally linked to this event—but what forces are responsible for the emergent cholesterol? The root cause of human depression remains a mystery, and vague causal factors, such as genetics and environment, are of little help.

Here is one speculative possibility. Around three million years passed between the time of the hominid Lucy and the advent of modern *Homo sapiens*. During that time our ancestors' brains expanded in size by nearly 300 percent. Both archaeological evidence and comparative analyses with our closest living relatives, the great apes, indicate that our prefrontal cortex experienced a disproportionate increase in the number of neurons present. Intriguingly, clinical neuroscience has linked dysfunction of the prefrontal cortex with a variety of mental illnesses, including depression, bipolar disorder, schizo-

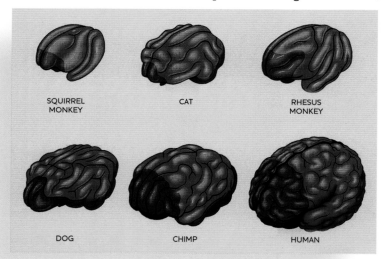

SQUIRREL MONKEY   CAT   RHESUS MONKEY

DOG   CHIMP   HUMAN

Comparison of prefrontal cortex in relation to overall brain size in mammals

phrenia, obsessive-compulsive disorder, and ADHD. This suggests that perhaps only animals with significant cortical development experience mental illnesses.

Our prefrontal cortex regulates a number of cognitive capacities, including attention, short-term memory, foresight, planning, and the ability to inhibit our more "reptilian" desires, such as lust and hunger (humans are probably the only species capable of voluntarily giving up sexual activity or consumption of food). Many neuroscientists also believe that our self-awareness, or reflective consciousness, is rooted in this mass of neuronal tissue. So human self-awareness probably evolved over a period of several million years, during which our hominid ancestors roamed the savannas of Africa and eventually the most distant corners of Europe and Asia. There is little doubt that life at this point in human evolution was exceedingly difficult. Survival was tenuous, infant mortality was high, and food and water were often scarce. Imagine now these same hominids becoming aware of the quality of their existence, no longer psychologically impervious to the harsh realities of life. As the cerebral cortex evolved across generations of men, tool use and language emerged, and perhaps a dawning recognition of our inevitable death. We might imagine one of our ancestors, sitting beneath an acacia tree, belly grumbling, legs sore, disease festering in a recent wound, pondering the meaning of her life. Might her newly developed cognitive capacities lead to the sobering conclusion that the only rational course of action was to lay down and die, so as to end the misery sooner? Such would be the maladaptive, paralytic nature of conscious self-reflection. This angst-ridden hominid would most likely fail at the dual tasks of survival and reproduction, which ultimately serve as the only currency of meaning within the world of natural selection.

To prevent existential paralysis, perhaps hominids evolved yet

In 2000, Prozac was listed as one of *Fortune* magazine's "Products of the Century."

Prozac
Installation view, Tang Museum

another mental adaptation: one that would allow them to grow and love and struggle and die without concern for the inherent meaninglessness of the mortal cycle. This cognitive shield could protect them from their own awareness or at least minimize the negative emotional impact such self-assessment might have on psychological well-being. We have inherited this shield from our successful ancestors, living without conscious awareness of its presence. It allows us to persevere through dull and seemingly pointless jobs, feel comfort in a daily routine, and find hope when the inevitable obstacles and tragedies of life materialize. For much of the world's population life remains bitterly harsh, filled with disease, war, and misery. Yet our species fights on and reproduces successfully, and most of us are surprisingly capable of living in ignorant bliss.

Unfortunately, no design is perfect, and our shield has the potential for failure. Let us hypothesize that this shield is biochemically maintained by neuronal release of serotonin into the prefrontal cortex. A serotonin deficiency, mediated in part by an unlucky genetic roll of the dice, would result in a weakened shield. The individual's psyche is slowly exposed to the horrors of reality and the futility of existence. A cluster of pathological, psychological, and behavioral symptoms emerges that we label depression. And what does Prozac do? It reestablishes our shield (restores our rose-colored glasses) and alleviates our behavioral despair. The cylinder of water no longer appears quite as daunting: after all, things are bound to change eventually! It gives hope back to the patient, providing strength to tolerate the common setbacks of life that previously seemed unbearable. We, like the mice, swim on.

This possibility for transformation is what ultimately makes Prozac worthy of mystique. How odd that a simple pill can alter our personal relationship with the universe.

1990

BUCKYBALL & CARBON NANOTUBES

C_{60}

U.S. PATENT NUMBER: 6,723,279 B1
CAS REGISTRY NUMBER: 99685-96-8

Scanning tunneling microscopy has a way of conjuring up fanciful thoughts. While lecturing about C_{60} and the fullerenes recently,
I visited one of the leading labs using this technique and looked
at images of these hollow, nanoscopic geodesic domes of carbon,
one at a time. Seeing such pictures, it is easy to imagine a
whole new world down there on the chemist's atom-by-atom length
scale, and there are those who will view this world with the
eyes and ambitions of a molecular architect. . . . Are these the first
elementary building blocks of a new carbon-based technology?

—RICHARD SMALLEY, 1992

by Cyrus Mody

Some molecules matter because of where they are useful: nylon in our hosiery and fishing line, penicillin in our bloodstreams. Some molecules, such as DNA, can even upend our understanding of the world. But once in a while a molecule matters because it becomes a rallying point, a symbol for a movement, a token of discoveries and applications just around the corner.

Buckyball, or C_{60}, is one such molecule, justly famous in both the scientific and popular imaginations. In the scientific literature alone C_{60} and its close relatives, carbon nanotubes, have been cited thousands of times in twenty-some years. Moreover, buckyball (also known as buckminsterfullerene) and nanotubes have entered popular culture and become symbols of the cutting edge of high tech.

Yet it is still rare to encounter actual buckyballs and carbon nanotubes outside the laboratory. Aspirin, isooctane, polyethylene—most people have regular contact with these molecules, but unless you buy the latest high-tech clothing, skin-care products, and sports gear, you have probably never met C_{60}. These new substances could someday change the world, by delivering medicines exactly where they are needed

in the body or by enabling faster, smaller transistors, solar cells, or space elevators.[1] But today that world-changing potential is still far off. Even if none of those applications come to pass, however, these would be molecules that matter. To understand why, we have to step—quite literally—off to the sidelines.

Note the goofy grins in these photos. One group has just made the discovery that will lead to the Nobel Prize in Chemistry, and the other group has just learned that two of them have won the Nobel Prize in Physics. The two photos were taken in September 1985 and October 1986. Two things connect these photos: soccer and nanotechnology. Both teams of scientists are gathered around a humble European football, and all of the Nobel laureates in these photos are celebrated pioneers of nanotechnology, the science of the very small.[2] Soccer and science are not an obvious pairing, but most new ideas come about through unusual combinations.

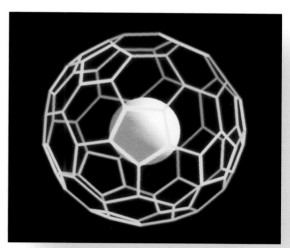

In the summer of 1978 a young, iconoclastic physicist, Gerd Binnig, was working at the IBM research lab near Zurich on something his colleagues said was impossible: a way to bring a sharp metal probe within a few atoms of a surface to watch electrons quantum-mechanically "tunnel." His aim was to move the probe back and forth over the surface like the electron beam on a television to see how the stream of electrons varied with the height of the surface. That way, he believed he would build a picture of a surface, right down to the individual atoms. The probe would act like a blind person's finger reading Braille, albeit in this case atom-sized Braille. Within four years Binnig's scanning tunneling microscope (STM) was routinely imaging atoms of silicon, graphite, and a few other metals and semiconductors. He had become a celebrity. Physicists crammed packed lecture halls to hear his results, and rumors of a Nobel Prize swirled around him.

By 1985 Binnig was sick of it. He was a maverick, an artist and a musician. Legend has it that on the first night his STM imaged individual atoms he became physically ill with excitement, then morose and despondent at the possibility of never discovering anything as wondrous again. So from 1985 to 1986 an exhausted Binnig took a sabbatical at Stanford University, where he joined Calvin Quate's group to relax and think about other things. Quate, a professor of

•

OPPOSITE TOP Left to right: Sean O'Brien, Richard Smalley, Robert Curl (standing), Harold Kroto, and James Heath
OPPOSITE BOTTOM Heinrich Rohrer (with flowers, left) and Gerd Binnig (with flowers, right), winners of the 1986 Nobel Prize in Physics for their invention of the scanning tunneling microscope, with the IBM Zurich research laboratory soccer team
ABOVE Computer simulation of the "shrink-wrap experiment," a metal atom enclosed in a C_{60} molecule

electrical engineering and applied physics, had organized a research group that studied applications of scanning probes. After a couple of months Binnig was lying on a couch in Quate's lab, staring at the ceiling, when his eyes opened to a new possibility. An STM needed a hard, sharp metal probe and only worked if the sample were a metal or semiconductor. But what if the probe were flexible, like a Slinky? It could be placed near a sample that would pull it down or push it away: either way, the amount of bend in the probe would indicate the height of the sample. It could be scanned to make a picture like the STM did, but now it could also look at samples that did not conduct electricity.

Binnig's eureka moment on the couch led to the atomic force microscope (AFM), one of the central instruments of nanotechnology. Today AFMs are used in the laboratories of physicists, chemists, materials scientists, geologists, biologists, biochemists, and many other disciplines; million-dollar AFMs are on the factory floor in fabrication facilities where computer chips are made; one AFM even crash-landed on Mars. STMs and AFMs allow us to interact with the nanoscale like never before: to push atoms around, snip molecules in two, or grab one end of a protein and yank on it until it straightens out. We do not have to deal with molecules in large numbers anymore because these microscopes allow us to think of single molecules as objects, even as tools.

What do Binnig, Heinrich Rohrer (codiscoverer of the STM), and their probe microscopy have to do with fullerenes? Probe microscopes and fullerenes are not intrinsically linked: you can look at fullerenes with many types of microscopes. The discoverers of C_{60} would probably have won their Nobel prizes without probe microscopy, but it would have taken longer for buckyball to matter. STMs and AFMs succeeded in making the nanoscale tangible. Buckyballs

Although scientists and researchers create fullerenes in the lab, they also occur naturally in outer space.

Melissa Gwyn
Buckminsterfullerene, 2007
Oil on wood panel
34 x 20 inches

and nanotubes were proclaimed as tiny toys: soccer balls to kick around with STM tips and Lincoln logs to assemble with AFM probes.

We have a good idea of how little splash fullerenes would have made if they had been discovered before probe microscopy. Fullerenes were made, seen, and talked about long before C_{60} was discovered in 1985, but no one paid them any attention. In fact, the best known early suggestion for a variant of buckyball appeared in the British journal *New Scientist* in 1966. The journal employed David Jones, an organic chemist, to write up fanciful scientific research under the pseudonym Daedalus—a prolific inventor employed by the fictional DREADCO. Daedalus explored the characteristics of a molecule formed from a graphite sheet curled up into a sphere that could trap smaller

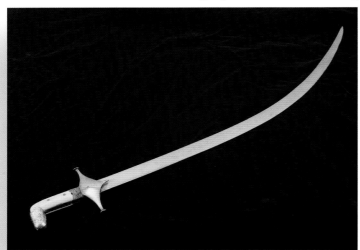

molecules in the interior—an idea that later proved crucial in demonstrating the existence of buckyball. However, because of the humorous nature of the Daedalus article Jones's suggestions were unlikely to be taken up. More serious explorations were equally neglected. For instance, in 1965 Harry Schultz, an organic chemist at the University of Miami, published a cursory description of geometrical solids, including the truncated icosahedron (or soccer ball), that could be formed by molecules of the type C_nH_n. Five years later Eiji Osawa, an organic chemist at Hokkaido University in Japan, published a paper on superaromaticity that included a detailed explication of soccer-ball C_{60}. Osawa's description, like Schultz's, never gained currency in the international

•
Shamshir (saber), Persian with Turkish decorations, c. 1880s
This saber was made from Damascus steel, which was coveted as early as the 17th century for its strength, flexibility, and light weight. Researchers have recently determined that these properties were present because the steel contains carbon nanotubes.

chemistry community, probably because his ideas were poorly abstracted in English and because there seemed little hope of synthesizing the molecule.

Both Schultz and Osawa came out of a tradition that would have been the natural point of origin for fullerenes. The team that eventually identified C_{60} had virtually no expertise with oddly shaped organic molecules that, like buckyball, resemble ordinary macroscale objects. Osawa, Schultz, and other synthetic organic chemists, however, had by the 1960s become obsessed with molecules that mimic ordinary objects: basketane, housane, pagodane, ovalene, cubane, snoutene, pterodactyladiene, birdcage, apolloane (resembling, loosely, the Saturn V rocket), lepidopterene (the "butterfly" molecule), and many others.

A human hair is as wide as a row of 80,000 buckyballs.

Most of these are whimsical names imposed on otherwise unexceptional molecules. Others were sought after as benchmarks of synthetic virtuosity, prized for their abstract aesthetics. The synthesis of cubane, for example, almost certainly triggered both Schultz's and Osawa's interest in truncated icosahedrons.

A few of these molecules are capable of caging smaller molecules within their frame, and some are components of molecular turnstiles, mechanical molecules that could have applications in everything from drug delivery to computing. The C_{60} soccer-ball molecule fit nicely into all these pigeonholes. It was a whimsical microscopic version of an ordinary object, a molecular analog of a pleasing geometrical shape, and a mechanical molecule capable of caging smaller molecules.

No wonder that the wheel (or in this case the soccer ball) kept getting reinvented. The soccer-ball shape was so intrinsically desirable that theoretical descriptions of it, and attempts to synthesize it, repeatedly cropped up. A Soviet team, D. A. Bochvar and E. G. Gal'pern, were next in 1973, followed by Robert Davidson in 1981, Josep Castells and Felix Serratosa in 1983, and Anthony Haymet in 1985. Castells and Serratosa even gave the molecule its IUPAC name: hentriaconta-cyclo[29.29.0.^{02,14}.0^{3,29}.0^{4,27}.0^{5,13}.0^{6,25}.0^{7,12}.0^{8,23}.0^{9,21}.0^{10,18}.0^{11,16}.0^{15,60}.0^{17,58}.0^{19,56} .0^{20,54}.0^{22,52}.0^{24,50}.0^{26,49}.0^{28,47}.0^{30,45}.0^{32,44}.0^{33,59}.0^{34,57}.0^{35,43}.0^{36,55}.0^{37,42}.0^{38,53}.0^{39,51} .0^{40,48}.0^{41,46}]hexacontane!

There are basically two routes to a new molecule: you can hypothesize it and then synthesize it, or you can stumble on it and then analyze what you have tripped over. Until 1985 everyone who worked on C_{60} took the first route—unsuccessfully. Orville Chapman at the University of California, Los Angeles, for example, set several generations of graduate students to futile attempts to synthesize "soccherene [*sic*] (I_h-C_{60})." The great insight of the team to win the Nobel Prize was that soccer-ball molecules could occur naturally. Other people before them had stumbled over C_{60} but had not connected their odd data to soccer balls and truncated icosahedrons.

The most visually stunning of these nondiscoveries was made by Sumio Iijima in 1980. Iijima, a virtuosic electron microscopist at Arizona State University, was looking at amorphous carbon samples when he encountered several beautiful, onion-shaped formations. They looked like order in a sea of chaos. He had no explanation for the formations, and his published images of them triggered little interest. Yet Iijima's career testifies to the ability of intriguing data to hibernate for years. Although he had long abandoned carbon, when he heard about buckyball in 1985, Iijima immediately made the connection to those old results. This time he knew what to look for: C_{60}, but also, he surmised, more esoteric forms of carbon. And after years of experimenting Iijima struck gold, discovering both the multiwalled (in 1991) and single-walled (in 1993) carbon nanotubes, cousins of buckyball that today show promise for use in everything from space elevators to molecular computing.

The name *buckminsterfullerene* is derived from R. Buckminster Fuller, the architect who invented geodesic domes.

In a more ironic twist C_{60} was "not discovered" one last time, by a group using an instrument similar to the one that would a few months later enable the "discovery" of C_{60}. In the late 1970s a young chemistry professor named Richard Smalley built (with his students) a laser supersonic cluster beam apparatus dubbed the AP2. It could zap samples with a laser to create small atomic clusters, inject those clusters into a carrier gas (such as helium), expand the gas to cool the clusters, ionize them with a second laser, and then examine them spectroscopically. Smalley began his career at Shell Chemical, developing ways to characterize polyolefins, an important ingredient in most plastics. Throughout his career Smalley would look back to the New Jersey petrochemical industry for ideas and collaboration.

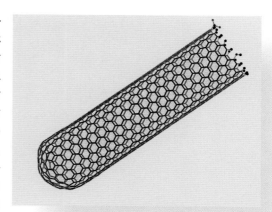

One of Smalley's partners was Andrew Kaldor at Exxon. For several years Smalley ran Kaldor's samples of uranium on the AP2 and shared the results with Exxon. Eventually, Kaldor wanted to run his own samples, so Exxon paid Smalley to build a variant of the AP2 (called the AP3) and send a copy to New Jersey. By 1984 Kaldor had this instrument up and running, but the American nuclear power industry had not kept pace. Kaldor's research on uranium therefore wound down, and he turned his attention to carbon, the main constituent of oil. Kaldor was motivated as much by grand scientific curiosity as by Exxon's bottom line. Knowing more about carbon clusters might yield a better oil-cracking method, but it could also reveal something awe inspiring about our place in the cosmos.

By 1970 it had become obvious that the heavens contained large, complex carbon molecules. Radio telescopy showed long carbon chains, such as HC_9N, in interstellar dust clouds, as well as a class

Diagram of a partial single-walled carbon nanotube

Buckyball and carbon nanotubes
Installation view, Tang Museum
FOREGROUND Richard Smalley's
laboratory notebook (contemporary
digital reproduction) and AP2 parts

of cosmic spectroscopic signatures not associated with any earthly molecules. Kaldor's group surmised that by zapping graphite they might learn something technologically useful, but by simulating interstellar conditions they would almost certainly be learning something about the ingredients of outer space and possibly about the origins of life.

So they put graphite into the AP3 and blew the carbon clusters into a time-of-flight mass spectrometer, an instrument that shows how massive a cluster is by measuring how long it takes to accelerate toward a detector. Oddly, a few very small clusters appeared, mostly containing from ten to twenty-three carbon atoms. The number of small clusters tailed off around thirty carbons; but there was a class of larger clusters, with the amount of carbons beginning around thirty-eight, with numbers peaking between fifty-six and seventy-six, then tailing off at a hundred.

Even more surprising, the larger clusters contained only even numbers of carbons: there was virtually no detection of clusters with sixty-seven or sixty-nine carbon atoms, but there were large signals at sixty-six or sixty-eight. Seemingly less significant were the noticeable peak at sixty carbons and the smaller one at seventy. There was still a lot of black art used to make the AP3 run well and interpret its results, so it was easy to dismiss these peaks as outliers on an otherwise smooth distribution. One last time buckyball was seen but not discovered.

Meanwhile, Harold Kroto, an astrochemist at the University of Sussex, had been pestering Smalley for months to conduct essentially the same experiment. Smalley had been introduced to Kroto by a fellow spectroscopist at Rice University, Robert Curl, in early 1984. Ever since, Kroto had tried to convince Smalley to see the AP2 as a simulator for the conditions in a dying star: great heat (as the star turns the last of its hydrogen and helium fuel into heavier ele-

A man's beard grows one nanometer in the time it takes him to lift a razor to his face.

ments) followed by great cold as the star dies, expands, and ejects a shell of those elements into the void. Kroto was convinced the AP2 would yield new molecules that could be sought for in the heavens. But Smalley—especially once he learned that his Exxon colleagues had already done the experiments—was lukewarm. Why switch to carbon when the AP2 was working on more technologically relevant metal and semiconductor clusters?

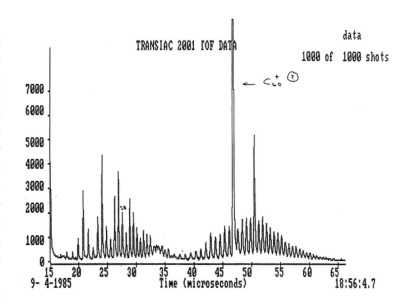

Finally, in late August 1985 Smalley had an opening in his schedule, and Kroto hurried to Houston. He parked himself beside the AP2 and its graduate student keepers, James Heath and Sean O'Brien.[3] All three excitedly strategized operating conditions, tinkered with equipment, and waited impatiently for the time-of-flight statistics to appear. Unsurprisingly, those statistics looked a great deal like Exxon's, except that on some runs the carbon peaks of sixty and seventy atoms were much larger than in the Exxon data. O'Brien and Heath found over ensuing days that they could tune the degree to which the peak at sixty dominated, up to more than forty times the frequency of its nearest neighbors.

This tunability showed the C_{60} peak was more than an experimental outlier: something made a cluster of sixty—and to a lesser extent seventy—carbon atoms more stable than other configurations. Why sixty? During a frantic couple of days Kroto, Curl, Smalley, O'Brien, and Heath tried to understand how graphite's hexagonal carbon rings

Time-of-flight mass spectroscopy experiment
by James Heath, September 4, 1985

could reassemble in three dimensions. They knew they were looking for some kind of geodesic dome and so were already considering calling it "buckminsterfullerene." But each time they made a dome of hexagons (including a gummy-bear-and-toothpick model made by Heath and his wife and consumed by another student), the structure was unstable.

Had any of them played soccer the answer might have been obvious. As it was, a cardboard star map Kroto built for his children provided the answer. He mentioned in passing that he thought the star map contained both hexagons and pentagons. In the wee hours of the next morning Smalley began cutting out paper hexagons and pentagons, taping them together until they formed a stable structure with sixty vertices! The next morning he called the chair of Rice University's math department to ask what esoteric geometric object this was—and heard the sarcastic reply: "What you've got there, boys, is a soccer ball."

Today, soccer-ball C_{60} is so well known that it seems inconceivable the AP2 experiments could be interpreted any other way. Yet early on there was tremendous doubt that the Sussex-Rice team had identified the correct structure, much less discovered a new allotrope of carbon. The Exxon team in particular raised pointed questions. The Rice and Sussex groups offered indirect answers, but by 1989 their inability to offer compelling proof of the existence of buckyball meant the number of articles on the molecule actually began to decline. It looked like C_{60} might not matter after all.

Three things rescued fullerenes. First, in early 1990 Wolfgang Krätschmer and Kosta Fostiropolous made and identified macroscopic quantities of C_{60} in soot. This discovery enabled clear spectroscopic and diffraction signatures showing that buckyball was indeed shaped

Early paper model of the C_{60} molecule made by Richard Smalley. Black lines (representing double bonds) were added by Robert Curl, suggesting that the molecule could be aromatic.

like a soccer ball. Intriguingly, their method of synthesis of C_{60} closely resembled that used in Thomas Edison's carbon-filament light bulb. Soon after, Iijima announced his discovery of the carbon nanotube, and research in this area began to grow dramatically, along with the applications to which they might be relevant.

Finally, fullerene research intersected with probe microscopy. The first STM images of buckyball dramatically showed a layer of soccer balls with a few taller rugby-ball shapes of C_{70} sticking up. Fullerenes had been seen with electron microscopes in 1980, of course, but those instruments flattened out the molecule's structure. STMs and AFMs could show fullerenes in three dimensions, as round balls on a surface.

More than that, the STM or AFM tip could nudge them around. This ability became crucial once nanotubes were discovered. It was clear that nanotubes would have unusual electronic characteristics: indeed a tube's chirality, or twist, makes it a semiconductor, a conductor, or even a superconductor. But to test its electronics you have to hook it up to electrodes. You could just shake a bunch of nanotubes down onto some premade electrodes and test the ones that happen to lie at the right angle and position. But if you are making a sophisticated device, you need the nanotube to be in exactly the right spot, which you can achieve if you can drag it into place with an AFM.

Scientists began to see that the electronic properties of a nanotube depend on its shape. Much like what happens to water trying to run through a kinked garden hose, electrons balk at running through a kinked nanotube. And while nanotubes bend gracefully when placed on a surface, if you want to tie them in knots you need an AFM. At the end of the 1990s people realized that imaging with a normal AFM tip is exceedingly clumsy work. So they began gluing nanotubes to AFM probes, which helped imaging, especially on a

Using nanotechnology, it may someday be possible to fabricate machines on a scale currently unimaginable. These machines would perform tasks at the molecular level.

surface with deep trenches (like a computer chip) or soft membranes (like a living cell).

Together, buckyball and probe microscopes provide us with new objects of knowledge and new ways to know them. Neither one is exciting without the other, but both have mattered equally for how they have changed the way we organize scientific research and think about how knowledge should be created. As the quote that opens this essay shows, some people—especially Smalley—saw the STM and the AFM as enabling one to think of fullerenes as building blocks and therefore to think of the nanoscale as a place for engineering and architecture.

Just when fullerenes were first viewed by an STM, a new movement sparked the scientific and popular consciousness: nanotechnology. Smalley himself was a prophet of nanotech: the founder and leader, until his death in 2005, of the Smalley Institute at Rice University and a frequent guest at congressional hearings, on pop-science television programs, and at scientific prize ceremonies. And each time he presented his molecule, buckyball, as the emblem of the new frontiers of science and technology made possible by thinking small. Smalley and buckyball were crucial to making serious nanoscience research exist today.

So if you never meet a fullerene and even if the grand hopes for nanotubes never come true, these are molecules that matter. Seeing new connections between different fields has always been a crucial driver of innovation. Fullerenes did not matter until those connections could be realized, which required the invention of such new tools as the AFM. Now that we have these connections and can call them nanotechnology, we are sure to see another century of innovations that make our lives faster, freer, and more complicated.

1. Because C_{60} is shaped like a geodesic dome (and a soccer ball), the discoverers decided to name the molecule after R. Buckminster Fuller, inventor of that architectural structure. Since Fuller's nickname was Bucky, C_{60} became affectionately known as "buckyball."

2. Fullerenes are a family of all-carbon hollow molecular cages. Buckyball is a hollow, spherical fullerene with sixty carbon atoms. Carbon nanotubes are long, thin molecules made up of half a fullerene on each end, separated by a rolled-up sheet of graphite.

3. The National Nanotechnology Initiative says that nanotechnology is "the understanding and control of matter at dimensions of roughly 1 to 100 nanometers, where unique phenomena enable novel applications." Richard Smalley's definition is pithier: "Nanotechnology is the art and science of making stuff on the nanometer scale." A nanometer is a billionth of a meter—about as long as ten hydrogen atoms end to end.

4. Two other graduate students, Yuan Liu and Qing-Ling Zhang, also kept the AP2 going during this crucial period. Liu in particular finessed the AP2's software just before the momentous discovery.

Buckyball
Installation view, Tang Museum

MOLECULE COLOR KEY

This color key identifies the chemical elements as represented in the large molecular models that appear in *Molecules That Matter*, in the molecular renderings at the beginning of each essay, and in the dots at the top of each page alongside essay titles.

● carbon

● hydrogen

● nitrogen

● oxygen

● sulfur

● phosphorus

● chlorine

GLOSSARY

ABORTIFACIENT
A substance that induces abortion, the removal or expulsion of an embryo or fetus from the uterus, either resulting in or caused by its death.

ACTION POTENTIALS
A wave of electrical discharge that travels along the cell membrane. Action potentials are an essential feature of cellular life, rapidly carrying information within and between tissues.

ALPHA HELICAL
(also α−helix)
Secondary structure of a protein that is the specific folding of the polypeptide chain due to the formation of hydrogen bonds between carbonyl oxygen and amide nitrogen atoms. When the hydrogen bonds set within a protein chain, it can form an alpha-helical structure.

ATOMIC FORCE MICROSCOPE (AFM)
A type of scanning probe microscope that can achieve atomic-scale resolution by monitoring the deflection of a cantilevered tip, one atom in width, as it moves across the surface of a material making contact or near contact with nanometer-scale features on that surface.

AXON TERMINAL
A long, slender projection of a nerve cell, or neuron, that conducts electrical impulses away from the neuron's cell body, or soma.

BIOMAGNIFICATION
The increase in concentration of an element or compound, such as the pesticide DDT, that occurs in a food chain as a consequence of small organisms taking up and storing toxins from the environment and then being eaten by predators, who are in turn eaten by even larger predators, thus accumulating very high levels of the toxin. Biomagnification can also be caused by the lack of, or very slow excretion/degradation of the substance.

BIOMOLECULES
Chemical compounds that naturally occur in living organisms. Biomolecules consist primarily of carbon and hydrogen, along with nitrogen, oxygen, phosphorus, and sulfur. Other elements sometimes are incorporated but are much less common.

CHARGAFF'S RULE
Named for Austrian chemist Erwin Chargaff, the rule states that in any sample of DNA, the percentage of base A (adenine) is always equal to that of base T (thymine); and the percentage of base G (guanine) is always equal to that of base C (cytosine).

CHLORINATED HYDROCARBONS
Compounds containing carbon and hydrogen and at least one covalently bonded chlorine atom.

CHROMATOGRAPHY
A procedure based on physical absorption principles used for separating various components of a mixture of chemical substances. In its broad interpretation, chromatography is a combination of separation, identification, and quantitative measurements of a mixture.

CORPUS-LUTEUM HORMONES
(also known as gestagens)
A group of female sex hormones derived from pregnane, the parent compound of steroids. Progesterone, one of the natural hormones, is formed by the corpus luteum in the second half of the menstrual cycle and by the placenta during pregnancy.

ENDOCRINE DISRUPTER
Substances that disrupt hormone secretion from endocrine glands or disrupt the effect of hormones on target cells.

ENDOMETRIOSIS
A common medical condition in women of reproductive age, where the tissue that lines the uterus is growing outside the uterus, on (or in) other areas of the body. Normally, the endometrium is shed each month during the menstrual cycle; however, in endometriosis, the misplaced endometrium is usually unable to exit the body.

ENZYME
A protein catalyst that lowers the activation energy to speed up biochemical reactions in living cells.

FULLERENES
Named after Buckminster Fuller, an architect who designed polyhedral domes based on hexagonal and pentagonal faces. Fullerenes are large molecules of pure carbon in the form of hollow spheres, formed when graphite is sublimed in a helium atmosphere, but they are also found in nature. Buckyball is the best-known fullerene.

GENE SEQUENCING
Determination of the order in which nucleic acids (A, T, G, C) are arranged in a gene.

GENETICALLY MODIFIED CROPS
Foodstuffs that have had their genomes altered through genetic engineering.

GENOME
An organism's entire hereditary information that is located in the DNA. The genome is a complete DNA sequence of one set of chromosomes.

MOLECULAR STRUCTURE
The order of atomic connections that constitute a molecule. It determines many properties of a substance, including its chemical reactivity, polarity, phase of matter, color, magnetism, and biological activity.

NEUROTRANSMITTERS
Compounds released from presynaptic nerve endings into the synaptic cleft; after diffusion across the synapse they transmit information to a postsynaptic cell. The postsynaptic nerve ending has an outer membrane provided with receptors; the postsynaptic cell can be a nerve, muscle, or gland cell.

NEURONS
Electrically excitable cells in the nervous system that process and transmit information.

PHARMACOTHERAPY
The practice of treating diseases with drugs.

POLYMERASE CHAIN REACTION
Polymerase chain reaction (PCR) is a technique that is used to amplify the number of copies of a specific region of DNA in order to produce enough DNA to be adequately tested. This technique can be used to identify, with very high probability, disease-causing viruses and bacteria, a deceased person, or a criminal suspect.

PREFRONTAL CORTEX
The anterior part of the frontal lobes of the brain, lying in front of the motor and premotor areas.

PROTEIN
A polymeric natural material composed of amino acids, which carry out many metabolic functions in the cell. All enzymes are proteins.

PSYCHOTROPIC DRUG
A chemical substance that acts primarily upon the central nervous system where it alters brain function, resulting in temporary changes in perception, mood, consciousness, and behavior.

PYRETHRINS
A natural insecticide made from the dried flower heads of *C. cinerariifolium* and *C. coccineum* used for insect control.

SEMICONDUCTORS
Any material that has specific electrical resistance under standard conditions and that is intermediate between those of conductors and insulators.

SEROTONIN
A vasoactive, biogenic amine (a biomolecule) that is a major neurotransmitter.

STAPHYLOCOCCI
Bacteria that can cause a wide variety of diseases in humans and other animals either through toxin production or invasion. For example, staphylococcal toxins are a common cause of food poisoning.

SUBLIMED/SUBLIMATION
The process by which a substance goes directly from the solid phase into the gas phase without passing through the liquid state.

SUPERCONDUCTOR
Any material that has very low or no electrical resistance. Such materials are being investigated to see if superconductivity can be achieved at, or near, room temperature. If such a superconductor can be found, electrical transmission lines with little or no resistance would be built, thus conserving energy lost in transmission and opening the way to many powerful technological applications.

SUPRAMOLECULAR
The area of chemistry that focuses on noncovalent bonding interactions of molecules. Supramolecular chemistry uses weak and reversible noncovalent interactions, such as hydrogen bonding, metal coordination, hydrophobic forces, van der Waals forces, pi-pi interactions, and electrostatic effects to assemble molecules into multimolecular complexes.

VASODILATION RESPONSE
The response that occurs when blood vessels in the body become wider following the relaxation of the smooth muscle in the vessel wall, thus reducing blood pressure.

X-RAY CRYSTALLOGRAPHY
When a pure crystal of a substance is irradiated by a beam of X-rays of a particular wavelength, the X-rays are scattered in different directions and can be detected using photographic film, scintillation detectors, or other devices. This defraction pattern is analyzed to determine the precise structure of the material (e.g., location of atoms in the crystal) causing the scattering.

CONTRIBUTORS

Mary Ellen Bowden, senior research fellow at the Chemical Heritage Foundation (CHF) in Philadelphia, Pennsylvania, has been associated with that institution since 1988. She holds a doctorate in the history of science and medicine from Yale University. She has curated several of CHF's exhibits and written extensively. In these endeavors she has occasionally written on the petrochemical industry and its icons.

Raymond J. Giguere was named Class of 1962 Term Professor of Chemistry at Skidmore College, Saratoga Springs, New York, in 2007. He joined the faculty at Skidmore in 1988 and holds a doctoral degree in organic chemistry from the University of Hannover, Germany, and a B.A. from Kalamazoo College in Michigan. In 1990 he created an interdisciplinary course for nonscience majors, "Playing Nature: Organic Synthesis and Society, 1900–1975," which he taught at Skidmore for nearly a decade. That pedagogical experience afforded him a conceptual basis for the *Molecules That Matter* exhibition.

Daniel R. Goldberg received his B.A. in chemistry from Skidmore College in 1992. He went on to obtain his Ph.D. in chemistry at Emory University and completed a postdoctoral program at the University of Chicago. Since 1998 Goldberg has worked as a medicinal chemist for Boehringer Ingelheim Pharmaceuticals in Ridgefield, Connecticut, where he continues to search for new therapies to treat inflammation and autoimmune diseases.

Neil Gussman is the communications manager for the Chemical Heritage Foundation. He is also news manager for *Chemical Heritage* magazine and writes a bimonthly column titled "We're History" for *Chemical Engineering Progress* magazine. He has owned a 1969 Ford Torino 428 Cobra Jet with Ram Air, a factory Holley carburetor and Hurst shifter. He has also owned a 1972 Ford Mustang 351 Cobra Jet Fastback.

Robert J. Hargrove has been a professor of organic chemistry and environmental science at Mercer University in Macon, Georgia, since 1975. In 1981, as a Fulbright Lecturer in Liberia he became interested in environmental health issues, including the DDT and malaria problem. He earned a master's degree in public health with a concentration in environmental and international health in 1987. Hargrove has worked in or visited West and East Africa. He is currently a volunteer professor one semester each year at Cuttington University, Suakoko, Liberia, West Africa.

Hassan López is an assistant professor in the Department of Psychology and Neuroscience Program at Skidmore College. He received his doctorate in behavioral neuroscience from the University of California at Santa Barbara. His scholarly interests center on the neurobiological basis of sexual attraction, desire, and courtship behavior, using both rodent and human experimental models.

Mary C. Lynn is the Douglas Family Professor of American Culture at Skidmore College. She earned her B.A. at Elmira College and her doctorate at the University of Rochester. She joined the faculty of Skidmore in 1969 and the American Studies department in 1970. Her research interests include the changing roles of women in the twentieth century and the military history of the Revolutionary War. She is the author of *Make No Small Plans: A History of Skidmore College*. Lynn currently teaches "Born in America," a cultural history of sexual reproduction in the United States.

Jeffrey L. Meikle is professor of American studies and art history at the University of Texas at Austin. His interests as a cultural historian include industrial design and technology, visual representation in popular print media, and alternative cultures from 1960 to the present. He is the author of *Design in the U.S.A., American Plastic:*

A Cultural History, and *Twentieth Century Limited: Industrial Design in America, 1925–1939*.

Cyrus Mody is an associate professor in the history department at Rice University in Houston, Texas. He received an A.B. in mechanical and materials engineering from Harvard University. His current research on the history of nanotechnology and corporate-academic relations in American science is an extension of his Ph.D. dissertation in science and technology studies at Cornell University.

Vasantha Narasimhan obtained her Ph.D in biochemistry from State University of New York at Albany in 1975. She joined the Department of Chemistry at Skidmore College in 1980 and served as the chair of the department from 2002 until her retirement in May 2006. The focus of her research for the last thirty-five years has been the study of structural modifications in DNA owing to its interaction with molecules such as chemotherapeutic agents of natural and synthetic origin and the biological implications of such interactions.

Audra J. Wolfe is editor in chief of *Chemical Heritage* at the Chemical Heritage Foundation in Philadelphia, Pennsylvania. She holds a Ph.D. in the history and sociology of science from the University of Pennsylvania and primarily researches the history and politics of the life sciences in the twentieth century.

CHECKLIST

Molecular models for *Molecules That Matter* were commissioned by the Frances Young Tang Teaching Museum and Art Gallery and the Chemical Heritage Foundation and were fabricated by Phil Fraley Productions, Inc. Unless otherwise noted, all objects are courtesy of the Frances Young Tang Teaching Museum and Art Gallery, Skidmore College.

INTRODUCTION

Vials of pure aspirin, isooctane, penicillin G, nylon, polyethylene, DNA, progestin, Prozac, and buckminsterfullerene

DDT is a controlled substance in the United States and is therefore not on display.

ASPIRIN

Aspirin molecule, 2006–2007
Anodized aluminum and polyurethane
84 x 84 x 48 inches

Fred Tomaselli
13,000, 1996
Aspirin, acrylic, and resin on wood panel
48 x 48 inches
Courtesy of the artist and James Cohan Gallery, New York City

Bayer aspirin advertising posters, 1938–1955
Contemporary digital reproductions
12½ x 18½ inches, each framed
Courtesy of Bayer AG/ Corporate History & Archives

Aspirin car, original 1930
Contemporary digital reproduction
5 x 7 inches
Courtesy of Bayer AG/ Corporate History & Archives

Bayer aspirin advertising poster, c. 1950
Advertising poster
13 x 10½ inches

"Almanaque," 1933
Mexican Bayer aspirin advertising poster and calendar
Ink on paper
16 x 27½ inches

Bayer aspirin dispenser, c. 1950
Metal
17½ x 8½ x 5½ inches

Bayer aspirin dispenser, c. 1930
Metal
13 x 3 (at center) x 3 (at center) inches

Certified Pharmacal Company aspirin dispenser and containers, c. 1940
Paper
12 x 8 x ½ inches
Courtesy of Raymond Giguere

Bayer aspirin advertising sign, c. 1920
Tin
15 x 18½ inches

Assorted Bayer aspirin containers, c. 1910–2007
Glass, tin, and plastic
Sizes vary

Aspirin Production Flow Sheet, date unknown
Contemporary digital reproduction
8 x 11 inches
Courtesy of Chemical Heritage Foundation Collections

Ball-and-stick models of aspirin, water, salicylic acid, and acetic acid, 2007
Plastic
Sizes vary

Aspirin "Chemistry Hall of Fame" drinking glass, c. 1970
Glass
2¼ x 3½ (diameter) inches

ISOOCTANE

Isooctane molecule, 2006–2007
Anodized aluminum and polyurethane
72 x 48 x 42 inches

Robert Dawson
Aerial View of Oildale, CA, 1991
Gelatin silver print
16 x 20 inches
Courtesy of the artist

Robert Dawson
Homes refinery, Long Beach, CA, 2006
Gelatin silver print
16 x 20 inches
Courtesy of the artist

Ed Ruscha
Mocha Standard, 1969
Color screen print on mold-made paper
24⅞ x 40 inches
Collection of Brook Alexander Editions (Tang installation)
Collection of Jordan D. Schnitzer, Portland, Oregon (all other venues)

The Getaway: A Collection of Car Chases, 1916–Present, 2007
DVD

Toyota engine, c. 1995
Metal
41 x 23 x 24½ inches
Courtesy of Bob Ensign, Ensign Auto Body

Chevrolet Engine, 1929
Metal
42 x 20 x 35½ inches
Courtesy of Diane and Howard Kloss

Ethyl gasoline pump, c. 1940
Metal and glass
24 (diameter) x 74 x 17 inches
Courtesy of James W. Tribley

"Right, Son—Change Now for Summer," c. 1950–1960
Mobil Oil and Gasoline advertisement
Ink on paper
15 x 11½ inches framed

"You Can Make Your Gas Go Further," c. 1950–1960
Mobil Oil and Gasoline advertisement
Ink on paper
14¾ x 11½ inches framed

"Flying Horse Power," c. 1950–1960
Mobil Oil and Gasoline advertisement
Ink on paper
15 x 20⅞ inches framed

Pegasus, c. 1935
Mobil gas station sign
36 x 48 inches
Courtesy of Raymond Giguere

Mobil Oil container, c. 1935
Glass and metal
14 x 3 x 3 inches
Courtesy of Joan M. Stenerson

Garford Motor Car, 1911
Magazine advertisement
Ink on newsprint
14¾ x 10½ inches

Chevelle, 1966
Magazine advertisement
Ink on paper
14 x 21 inches

Ford Mustang and Mach 1, 1970
Magazine advertisement
Ink on paper
12¼ x 17½ inches

Dodge Charger, 1971
Magazine advertisement
Ink on paper
14 x 21 inches

Oil barrel, 2006
42-gallon drum, 34½ x 23 (diameter) inches
Courtesy of Arrow Transport, Inc.

PENICILLIN

Penicillin molecule,
2006–2007
Anodized aluminum and
polyurethane
129 x 50 x 72 inches

Frank Moore
Beacon, 2001
Oil on canvas over featherboard
72 x 96 inches
Courtesy of the Joy of Giving
Something Inc., New York City

Jean Shin
Chemical Balance 2, 2005
Prescription pill bottles,
mirrors, and epoxy
7 units from 32 to 46 inches
in diameter;
overall dimensions variable
Courtesy of the artist
and Frederieke Taylor Gallery,
New York City

Models of penicillin G,
penicillin F, amoxicillin,
and ampicillin, 2007
Plastic
Sizes vary
Courtesy Indigo Instruments

Penicillin bottle, 1949
Abbott Laboratory
Glass
3½ x 5 inches
Courtesy of Chemical Heritage
Foundation Collections

Image of penicillin
fermentation bottles,
c. 1945
Contemporary digital
reproduction
8 x 10 inches
Courtesy of Chemical Heritage
Foundation Collections

Alexander Fleming, n.d.
Contemporary digital
reproduction
8 x 10 inches
Courtesy of Chemical Heritage
Foundation Collections

"Hey boy friend, coming
MY way?," c. 1940
World War II
medical propaganda poster
25¼ x 17 inches

Alexander Fleming,
May 15, 1944
Time magazine cover
Ink on paper
10¾ x 8¼ inches

"Penicillin," May 24, 1943
Three-page excerpt from *Life*
magazine article
Ink on paper
14 x 10⅜ inches

"Thanks to Penicillin, He
Will Come Home!," c. 1944
Penicillin advertisement,
Schenley Laboratories, Inc.
Ink on paper
14 x 10⅜ inches

"Powder of Life," 1944
Penicillin advertisement,
York Refrigeration, Inc.
Ink on paper
10¾ x 8¼ inches

Glasbake penicillin
fermentation bottle, c. 1940
Glass
2 x 4 x 10¼ inches
Courtesy of Gayle Rettew King

"The Wonder of Penicillin," 1950
Magazine advertisement
Ink on paper
13¾ x 10½ inches

POLYETHYLENE

Polyethylene molecule,
2006–2007
Anodized aluminum and
polyurethane
108 x 24 x 30 inches

Tony Cragg
New Figuration, 1985
Plastic wall construction
113 x 54 inches
Collection Speed Art Museum

Kara Daving
*Aruba, Dutch Caribbean,
JE (Irausquin Boulevard)*,
May 23, 2005
*Buffalo, NY (Elmwood
Avenue)*, April 19, 2005
*Bowling Green,
OH (Kenwood Avenue)*,
April 10, 2006
*Pittsburgh, PA (Murray
Avenue)*, September 4, 2005
11 x 13 inches each
Photo transfers on plastic bags
Courtesy of the artist

Roxy Paine
S2-P2-BK1, 2006
S2-P2-BK2, 2006
S2-P2-BK10, 2006
S2-P2-BK15, 2006
S2-P2-BK18, 2006
S2-P2-BK23, 2006
Low-density polyethylene
28 x 22 x 24 inches
Courtesy of James Cohan Gallery

Dan Peterman
Tische, 1996
Recycled, post-consumer
plastic
15 x 44 x 19¾ inches
Courtesy of the artist and
Andrea Rosen Gallery

Ball-and-stick models of
polyethylene terephthalate
(PETE), high-density
polyethylene (HDPE),
polypropylene (PP),
low-density polyethylene
(LDPE), polyvinyl chloride
(PVC), polystyrene (PS),
2007
Sizes vary
Courtesy Indigo Instruments

Wham-O Frisbee, c. 1950
Polyethylene
8 (diameter) inches

Assorted Tupperware,
c. 1950–1970
Polyethylene
Sizes vary

Assorted toys
and Tupperware,
c. 1950–2007
Polyethylene
Sizes vary
Courtesy of Chris Daily

Raw polyethylene pellets
and remnant from
blow-molding process,
2007
Low-density polyethylene
Sizes vary
Gift of Eastern Packaging
Lawrence, Massachusetts

High- and low-density
polyethylene
containers and objects
20 x 20 x 96 inches
Donations collected by
Skidmore College Chemistry
Club

Don Featherstone Pink
Flamingos, 2006
Union Products, Inc.
High-density polyethylene
30 x 16 and 24 x 16 inches

NYLON

Nylon molecule, 2006–2007
Anodized aluminum and
polyurethane
142 x 48 x 78 inches

Susie Brandt
After Albers, 1995–1998
Nylon panty hose, handwoven
on a potholder loom
72 x 58¾ inches
Courtesy of the artist

Presentation on Joseph X.
Labovsky, 2005
Courtesy of Chemical Heritage
Foundation Collections

Six display legs with
stockings, 2007
Nylon and plastic on
wood board
31 x 60 x 24 inches

Metal forming leg
for preboarding nylon
stockings, 2007
Contemporary reproduction
44 x 7 inches
Courtesy of Chemical Heritage
Foundation Collections

Vintage stockings and
boxes, c. 1940
Nylon
(white) 9 5/8 x 7 1/4 x 3/4 inches,
(green) 9 3/4 x 7 1/8 x 1/2 inches,
and (light brown) 37 x 6 1/2
inches
Courtesy of Chemical Heritage
Foundation Collections

DuPont nylon stocking
advertisements, n.d.
Magazine advertisements
(stocking legs in closet),
11 1/4 x 8 3/4 inches,
and (stocking in fan shape)
12 1/2 x 8 7/8 inches
Courtesy of Chemical Heritage
Foundation Collections

Wallace Carothers' laboratory notebook, November 17,
1925–August 6, 1927
Ink on paper
8 1/4 x 6 7/8 x 1/2 inches
Courtesy of Chemical Heritage
Foundation Collections

Wallace Carothers in his
laboratory, c. 1925
Contemporary
digital reproduction
8 x 10 inches
Courtesy of Hagley Museum
and Library

The DuPont Magazine,
September 1941
Ink on paper
11 1/4 x 8 1/4 inches
Courtesy of Chemical Heritage
Foundation Collections

This Is DuPont, November 1952
Ink on paper
12 x 9 inches
Courtesy of Chemical Heritage
Foundation Collections

50th Anniversary of Nylon,
Special Report, April 1988
Ink on paper
11 x 8 inches
Courtesy of Chemical Heritage
Foundation Collections

Nylon, The First 25 Years,
1963
Ink on paper
11 3/4 x 8 3/4 inches
Courtesy of Chemical Heritage
Foundation Collections

*Better Living, DuPont
Employee Magazine: Nylon's
10th Anniversary*,
November–December, 1948
Ink on paper
12 x 9 inches
Courtesy of Chemical Heritage
Foundation Collections

Box of 12 Dr. West
toothbrushes, 1956
Polyethylene, nylon, and
cardboard
8 x 8 x 2 inches

Dr. West toothbrush in
glass tube, 1938
Polyethylene, glass, and nylon
3/4 (diameter) x 7 inches

Tire cord, 1944
Nylon
9 5/8 (height) x 5 (diameter)
inches
Courtesy of Chemical Heritage
Foundation Collections

Autoclave sample, 1937
Nylon
8 1/2 x 2 x 1/8 inches
Courtesy of Chemical Heritage
Foundation Collections

Demonstration of cold
drawing polyester, 1930
Contemporary digital
reproduction
8 x 10 inches
Courtesy of Hagley Museum
and Library

250-pound autoclave
and casting equipment,
November 8, 1937
Contemporary digital
reproduction
8 x 10 inches
Courtesy of Hagley Museum
and Library

Detail of BJ-4 drawtwister,
July 15, 1938
Contemporary digital
reproduction
8 x 10 inches
Courtesy of Hagley Museum
and Library

Yarn inspection, late 1938
Contemporary digital
reproduction
8 x 10 inches
Courtesy of Hagley Museum
and Library

Inspecting staple drawing
machine, February 1947
Contemporary digital
reproduction
8 x 10 inches
Courtesy of Hagley Museum
and Library

Joseph X. Labovsky
at DuPont's Seaford Plant,
1939
Contemporary digital
reproduction
8 x 10 inches
Courtesy of Hagley Museum
and Library

Nylon pirns awaiting final
inspection, n.d.
Contemporary digital
reproduction
8 x 10 inches
Courtesy of Hagley Museum
and Library

Dress, c. 1940s
Parachute nylon
52 1/2 (length) x 13 (at waist) x
16 (at bust) inches
Courtesy of Chemical Heritage
Foundation Collections

"Rythm Ride," 1950
B.F. Goodrich tire
cord advertisement
Ink on paper
11 x 14 inches

Three women proudly
show their nylons, c. 1950
Silver gelatin print
4 x 4 inches

Pirns, 2006
Nylon
Approx. 7 x 3 (diameter)
inches each

Handmade parlor guitar,
c.1910
Spruce, mahogany,
and Brazilian rosewood
with nylon strings
36 x 12 inches
Courtesy of Raymond Giguere

Augustine guitar
strings, 2007
Nylon
10 1/2 x 14 1/2 x 1 1/4 inches
Framed
Courtesy Raymond Giguere

DNA

DNA model, 2006
Plastic and wood
36 x 14 (diameter) inches
Courtesy Indigo Instruments

Bryan Crockett
Anger from the *Seven
Deadly Sins* series, 2001
Cultured marble
12 x 13 x 20 inches
Courtesy of the Joy
of Giving Something, Inc.,
New York City

Bryan Crockett
Gluttony from the *Seven
Deadly Sins* series, 2001
Cultured marble
7 7/8 x 6 1/4 x 12 inches
Courtesy of the Joy
of Giving Something, Inc.,
New York City

Bryan Crockett
Sloth from the *Seven Deadly Sins* series, 2001
Cultured marble
Approx. 18 x 11 x 18 inches
Courtesy of the artist

Michael Oatman
Code of Arms, 2004
Collage on printed paper with test-tube rack frame
106¾ x 54 inches
Courtesy of the artist

Alexis Rockman
Romantic Attachments, 2007
Oil and wax on wood panel
120 x 96 inches
Courtesy of the artist and
Leo Koenig, Inc.

Erwin Schrodinger
What Is Life?, 2006
Cambridge University Press,
1st published 1944
8½ x 5½ x ½ inches

Human Gene Chip
US Patent No: 5,744,305
and 5,445,934
Human Genome U95A
2¾ x 1½ inches
Courtesy of Bernard Possidente

NATURE: The Human Genome, February 2001
Magazine cover
10¾ x 8¼ inches
Courtesy of Bernard Possidente

SCIENCE: The Human Genome, February 2001
Magazine cover
10⅜ x 8¼ inches
Courtesy of Bernard Possidente

Human Gene Chip
U 113 Plus 2.0 P/N: 520019
2¾ x 1½ inches
Courtesy of Matthew Isakson
and Abbott Labs, Chicago

Mouse Gene Chip
430 2.0 Array P/N: 520029
2¾ x 1½ inches
Courtesy of Matthew Isakson
and Abbott Labs, Chicago

Pink or Blue, The Early Gender Home DNA Testing Kit, 2007
8 x 6 x 1¼ inches
Donation from Pink or
Blue DNA

Ancestry by
DNA Home Testing Kit, 2006
5½ x 8 inches
Courtesy of Ancestry DNA

Tim Junkin
Bloodsworth, 2005
Algonquin Books of
Chapel Hill
8 x 5 inches
Courtesy of Raymond Giguere

"50 Years of the Double Helix," April 2003
Scientific American magazine cover
Ink on paper
10¾ x 8¼ inches

DNA Interactive, 2003
Interactive DNA DVD
Courtesy of Cold Spring
Harbor Laboratory

James D. Watson with Andrew Berry
DNA, The Secret of Life, 2003
Alfred A. Knopf Publishing
9¼ x 7¾ inches

Van R. Potter
DNA Model Kit, 1959
Burgess Publishing Company
Paper and plastic
Gift of Maureen Morrow
and Jennifer Waldo
Courtesy of Raymond Giguere

Rosalind Franklin
Photo 51, original 1952
Contemporary
digital reproduction
8 x 10 inches
Cold Spring Harbor
Laboratory, Cold Spring
Harbor, New York

Watson and Crick Walking along the Backs, 1952 (original)
Contemporary digital reproduction
8 x 10 inches
Cold Spring Harbor
Laboratory, Cold Spring
Harbor, New York

Dolly the sheep and her first lamb, Bonnie, 1993 (original)
Contemporary
digital reproduction
5 x 7 inches
Courtesy of the Roslin
Institute, Edinburgh

PROGESTIN

Progestin molecule, 2006–2007
Anodized aluminum and polyurethane
120 x 68 x 82 inches

Chrissy Conant
Chrissy Caviar® 2001–02
(series of 12 jars)
AP Shown.
One human egg (per jar),
human tubal fluid,
liquid silicone, polyethylene,
glass, brass, UV laminated
label, refrigeration equipment
Jar dimensions: 3¾ (width)
x 4 (diameter) x 3½
(height) inches
Case dimensions: 17 (width)
x 15 (diameter) x 35
(height) inches
Courtesy of the artist
www.chrissycaviar.com

Chrissy Caviar® 2001–02
(edition of 50, 3/50 shown)
Giclée print, gold silkscreen on
100% Cotton Crane's
Museo paper
35 x 27 inches (unframed)
Courtesy of the artist

Chrissy Caviar®
Limited Edition Floaty™
Pen, 2001–2002
(edition of 1,000) 24 Shown: PVC
plastic, liquid silicone, aluminum,
ink, Plexiglas display stand
6 x ⅝ (diameter) inches
Courtesy of the artist

Making Chrissy Caviar®
2001–02 (edition of 12)
DVD run time 10 minutes
Monitor size variable
Courtesy of the artist

Melissa Gwyn
Progesterone, 1992
Ink, pencil, watercolor, gouache,
and pastel on paper
28 x 22 inches
Courtesy of A.G. Rosen

Ovulen pill sample, n.d.
4¼ (diameter) x ⅜ (high) inches
Courtesy of Chemical Heritage
Foundation Collections

Enovid pill sample, n.d.
2¾ diameter x ¼ x 5½ inches
Courtesy of Chemical Heritage
Foundation Collections

Enovid pill packet, n.d.
Paper and progestin
2¾ x 5½ x ¼ inches
Courtesy of Chemical Heritage
Foundation Collections

Unknown pill sample, n.d.
2½ (diameter) x ⅜ x 3⅞ inches
Courtesy of Chemical Heritage
Foundation Collections

Ovex Tab Stick dispenser,
c. 1968
¾ (diameter) x 3 inches
Collection of the Dittrick
Medical History Center, Case
Western Reserve University

Ortho "U-Press-It" Dialpack,
c. 1975
3 (diameter) x ¼ inches
Collection of the Dittrick Medical
History Center, Case
Western Reserve University

Noriday tablets, n.d.
½ x 3 x 3½ inches
Collection of the Dittrick
Medical History Center,
Case Western Reserve University

Ortho-Novum 3-cycle
Dialpack, c. 1966–1968
3½ (diameter) x ¼ inches
Collection of the Dittrick Medical
History Center, Case
Western Reserve University

"An Oppressed Majority
Demands Its Rights,"
January 19, 1970
Life magazine article
Ink on paper
13 x 20 x ⅛ inches

"The Pill," April 7, 1967
Time magazine cover
Ink on paper
10 x 17½ inches

"Birth Control: The Pill and
the Church," July 6, 1964
Newsweek magazine cover
Ink on paper
10 x 8 inches

"Boy and The Pill," April 1969
Eye magazine cover
Ink on paper
13 x 9½ inches

"The New Pill: Should You
Take It," October 1985
Ms. magazine cover
Ink on paper
10 x 8 inches

Loretta Lynn and
Back To The Country
The Pill, 1975
MCA Records
Long-playing record album
12¼ x 12½ inches

The Brothers-In-Law
*The Pill Administered
Orally by The Brothers-In-
Law*, 1970
Arc Sound Ltd.
Long-playing record album
12¼ x 12½ inches

Brain
Birth Control Increase,
1977
D. Hitchcock and B.C.P.
Long-playing record album
12¼ x 12½ inches

Margaret Sanger at
Planned Parenthood, 1922
Contemporary
digital reproduction
8 x 10 inches

Planned Parenthood
stationery envelope,
postmarked March 18, 1972
Ink on paper envelope, stamp
4⅛ x 9½ inches

Buttons, 2006
Metal and plastic
¾ (diameter) inches each

"To Hell with Birth Control,"
c. 1960s
Metal and plastic button
3½ (diameter) inches

DDT

DDT molecule, 2006–2007
Anodized aluminum
and polyurethane
110 x 72 x 66 inches

Robert Dawson
*200 tons of DDT buried
underwater, Santa Monica
Bay, CA*, 1989
Gelatin silver print
16 x 20 inches
Courtesy of the artist

Melissa Gwyn
DDT, 2007
Oil on wood panel
36 x 24 x 2 inches
Courtesy of the artist

Peregrine falcon, 1977
Mounted specimen
18 x 18 x 18 inches
Courtesy of New York State
Department of Environmental
Conservation

Peregrine falcon, n.d.
Mounted specimen
16 x 14 x 5 inches
Courtesy of the Pember
Museum of Natural History
(all other venues)

Rachel Carson
Silent Spring, 1962
First edition, Houghton
Mifflin Company
9 x 6 x 2 inches

Frank Graham, Jr.
Since Silent Spring, 1970
Houghton Mifflin Company
9 x 6 x 2 inches

DDT Sprayer, c. 1950
FLIT Esso Standard
Oil Company
Metal
8 x 8 x 2 inches

Insect spray with DDT,
c. 1960
Standard Oil Company
Metal
7 x 5 x 2 inches

Quick Action DDT
Gulfspray, c. 1960
Gulf Oil Corporation
Metal
7 x 5 x 2 inches

DDT insect spray, c. 1960
Bee Brand
Metal
7 x 5 x 2 inches

DDT sprayer, c. 1960
Metal
29 x 8¾ inches
Courtesy of Chemical Heritage
Foundation Collections

Gulfsprayer, c. 1950
Gulf Oil Corporation
Approx. 4 x 12 x 4 inches
Courtesy of Robert Hargrove

Earth Day Society pin,
c. 1965
Metal and plastic
2 (diameter) inches

Earth Day pin,
c. 1960
Metal and plastic
2 (diameter) inches

Neocide DDT,
c. 1940
French advertisement
15¾ x 11¾ inches

L'Insectoline DDT,
c. 1930
French advertisement
15¾ x 11¾ inches

"DDT: Weapon of Mass
Survival," 2006
Cotton T-shirt

"April 22 Earth Day," 2006
Cotton T-shirt

"Bug-a-boo" DDT, 1946
Magazine advertisement
14 x 10¾ inches
Courtesy of Chemical Heritage
Foundation Collections

PROZAC

Prozac molecule, 2006–2007
Anodized aluminum and
polyurethane
88 x 60 x 124 inches

Thomas Asmuth
Fluoxetine, 2006
60 x 40 x 20 inches
Courtesy of the artist

Melissa Gwyn
Prozac, 2007
Oil on wood panel
16 x 11 inches
Courtesy of the artist

Eli Lilly Pharmaceutical
promotional materials,
cup, clock, puzzle game,
note pad, pens, 1995–2006
Sizes vary

Consumer kitsch
belt, magnets, stickers,
buttons, 2000–2006
Sizes vary

Drew Dernavich
"Edgar Allen Prozac,"
November 6, 2006
The New Yorker
5 x 7 inches

"Chance Made us Sisters,
Prozac Made us Friends," 2005
Greeting card
6¾ x 4¾ inches
Courtesy of Chemical Heritage
Foundation Collections

"Lost in a Maze of
Bewildering Thoughts," 1979
Marketing brochure
7¾ x 7¾ inches closed,
7¾ x 24¼ inches open
Courtesy of Chemical Heritage
Foundation Collections

"Man, by His Very
Nature Must Live with
Anxiety," 1967
Marketing brochure
8 x 8 inches closed,
8 x 16 inches open
Courtesy of Chemical Heritage
Foundation Collections

Assorted Books on Prozac,
1989–2007

BUCKYBALL

Buckyball molecule,
2006–2007
Anodized aluminum and
polyurethane
81 x 81 x 81 inches

Melissa Gwyn
Buckminsterfullerene, 2007
Oil on wood panel
34 x 20 inches
Courtesy of the artist

Shamshir (saber), 1880s
Persian with Turkish decorations
Wootz steel, iron, ivory, silver,
brass, adhesive
13¾ x 44⅞ x 6½ inches in case
Courtesy of Higgins
Armory Museum, Worcester,
Massachusetts

Arc discharge chamber,
1998
Iron, carbon, and chromium alloy
17¾ (diameter) x 18 inches
Courtesy of Pulickel Ajayan's
Carbon Materials Group,
Rensselaer Polytechnic Institute

Carbon nanotubes, c. 2005
Contemporary digital
reproduction
10¼ x 7¾ x 1¼ inches
Pulickel M. Ajayan,
Henry Burlage Professor of
Engineering
Carbon Materials Group,
Rensselaer Polytechnic Institute
© 2007 Rensselaer Polytechnic
Institute. All rights reserved.

ADC for the AP2 Supersonic
Laser Vaporization
Cluster Beam Apparatus,
c. 1995
Stainless steel, ceramic,
and copper
10 (diameter) x 5¼ inches
Courtesy of Chemical Heritage
Foundation Collections

Flange for the AP2
Supersonic Laser
Vaporization Cluster Beam
Apparatus, c. 1995
Stainless steel and copper
10 (diameter) x 3 inches
Courtesy of Chemical Heritage
Foundation Collections

Actuator for the AP2
Supersonic Laser
Vaporization Cluster Beam
Apparatus, c. 1995
Steel, aluminum, and plastic
6½ (diameter) x10¼ x 9 inches
Courtesy of Chemical Heritage
Foundation Collections

Richard Smalley's AP2
laboratory notebook, 1985
Contemporary reproduction
9½ x 7½ x ⅝ inches
Courtesy of Chemical Heritage
Foundation Collections

Adult Zyvex Easton 3, CNT
baseball bat, 2006
Youth Zyvex Easton 3, CNT
baseball bat, 2006
Enhanced composite
Adult: 2½ x 30 inches;
youth: 2¼ x 20 inches
Courtesy of Easton Sports

Cannondale bike frame, 2007
Enhanced composite
24 x 38 x 6½ inches
Courtesy of Cannondale Bicycle
Corporation

Carbon nanotube golf
club, 2007
Enhanced composite
47 x 5½ x 5½ inches
Courtesy of Grafalloy, Inc.

Carbon nanotube golf
balls, 2007
2 x 5 x 5 inches

Babolat Teamline
tennis racket
Enhanced composite and
nylon strings
28 x 12 x 1½ inches
Courtesy of Wilson Sports

Hol-e Roller Dog Toys, 2006
Rubber
5, 6½, and 8 inches
Courtesy of John and Karen
Sciolino

Ball-and-stick molecule
model of diamond, c. 1950
Wood and metal
8½ x 8 x 8 inches
Courtesy of Skidmore College
Chemistry Department

Ball-and-stick molecule
model of graphite, c. 1950
Wood and metal
8½ x 8 x 8½ inches
Courtesy of Skidmore College
Chemistry Department

Buckyball model, 2006
Plastic
6 (diameter) inches
Courtesy of Skidmore College
Chemistry Department

Carbon nanotube
model, 2006
Plastic
5 x 30 inches
Courtesy of Skidmore College
Chemistry Department

Carbon nanotube
model, 2006
Plastic
6 x 30 inches
Courtesy of Skidmore College
Chemistry Department

Chicken-wire carbon
nanotube model, 1998
11 (diameter) x 70 inches
Courtesy of Pulickel Ajayan's
Carbon Materials Group,
Rensselaer Polytechnic Institute

Signed *Time* magazine
cover of Buckminster Fuller,
January 10, 1964
Ink on paper
18 x 12 x 1 3/8 inches
Courtesy of Phil Haggerty

U.S. Patent of Laminar
Geodesic Dome, 1981
Signed poster
22¼ x 28¼ x 1⅜ inches
Courtesy of Phil Haggerty

Richard Smalley with two
students, 1994
Contemporary digital
reproduction
10¼ x 7¼ x 1¼ inches
Courtesy of the Chemical
Heritage Foundation
Photo Courtesy of Tommy
LaVergne

Richard Smalley, Harold
Kroto, and Robert Curl, Jr.,
with Nobel Prize, 1996
Color photograph
10¼ x 7¼ x 1¼ inches
Courtesy of Chemical Heritage
Foundation Collections

ACKNOWLEDGMENTS

A major traveling exhibition project with a catalog and extensive online features is always a collective undertaking. In the case of *Molecules That Matter*, the surprisingly large number of contributors provides a reliable index of the overarching historical, scientific, and cultural reach of the topic. *Molecules That Matter*—an examination of ten organic molecules that shaped life in the twentieth century—is the brainchild of Raymond J. Giguere, the class of 1962 Term Professor of Chemistry at Skidmore College. Ray's energy, passion, and vision have driven this project from start to finish, and I thank him for his tireless work, good humor, flexibility, and constant advocacy as my co-curator. The Tang's founding director, Charlie Stainback, originally approved the show, and Ian Berry, the Tang's Associate Director and Susan Rabinowitz Malloy '45 Curator, shepherded it ably through its first stages. Artist Fred Wilson also offered early encouragement for the idea of an exhibition uniting chemistry and contemporary art during his Luce Fellowship residency from 2004 to 2006.

Creating *Molecules That Matter* has been a joint project undertaken with the Chemical Heritage Foundation (CHF), the nation's premier institution devoted to preserving and promoting the history of chemistry. Collaborating with CHF has been a privilege and a pleasure, and we salute the work of our CHF partners in this effort, including Miriam Schaefer, Mary Ellen Bowden, Shelley Wilks Geehr, Julie Conners, Rick Sherman, Jennifer Landry, Erin McLeary, Rosanne DiVernieri, Patricia Wieland, and former CHF staffers John Van Ness and Robert Hicks. Our particular thanks go to Marjorie Gapp, CHF's curator, for her work in planning the exhibition's installation at CHF's Hach Gallery, and to installation designer Keith Ragone. The Tang Museum and Skidmore College could never have accomplished this project alone, and we thank our colleagues at CHF for helping bring it to fruition.

Funding for *Molecules That Matter* has been provided by The Camille and Henry Dreyfus Foundation, whose crucial early support helped launch the exhibition, and by the Hach Scientific Foundation, Amgen, Sara Lubin Schupf '62, the Friends of the Tang, and donors to the Chemical Heritage Foundation. We thank all of them for their generosity and belief in this project.

A diverse Scientific Advisory Board helped winnow the many possible molecular candidates down to the ten selected for the show. Along with Giguere himself, its members include John Van Ness, Julian Adams, Mary Ellen Bowden, Brad Herberich, Daniel R. Goldberg, Robert Hargrove, Vasantha Narasimhanand, and David Yerow. We are proud and delighted to have had two eminent scientists and Nobel laureates, Roald Hoffmann and Dudley Herschbach, vet the board's selections.

At Skidmore, many faculty members contributed ideas and expertise, including Giguere's colleagues in chemistry, Steven Frey, Michelle Frey, Shannon Stitzel, Kara Cetto Bales, Judith Halstead, David Wos, as well as Cheryl Towers and Jennifer McCluen. Along with Stitzel, Halstead, and Bales, additional Skidmore faculty members contributed talks to a *Molecules That Matter* symposium held at Skidmore in spring 2006, including Lisa Aronson, art history; Lewis Rosengarten, music; Bernard Possidente, biology; Mary C. Lynn, American studies; Rik Scarce, sociology; Reginald Lilly, philosophy; Hassan Lopez, psychology and neuroscience; William Standish, physics; Daniel Hurwitz, mathematics and computer science; Rachel Roe-Dale, mathematics and computer science; Kris Szymborski, Scribner Library; and Richard Wilkinson of the University of Albany. Through his fundraising efforts, Skidmore's Director of Foundations and Corporate Relations, Barry Pritzker, also supported the exhibition from its inception, and Don and Jean Richards have been enthusiastic supporters from the start.

Skidmore students and alumni have played a major role in creating *Molecules That Matter*. We would like to thank Riley Abair, Lea Chiara, Jessica Horowitz, Michelle Julian, Jackie Koustmer, Erin Manning, Alexandra Reingold, Blair Stelle, Lilly Torrey, and Emily Weiss for their many valuable contributions to the online feature. Tang interns and workers Kacey Light and Justin Hirsch have also been integral on many fronts of the exhibition, catalog, and online feature (http://tang.skidmore.edu/pac/mtm/). A number of chemistry students took part in researching key topics for *Molecules That Matter*: Lauren Adams, Alyssa Bennett, Luke Ceo, Julia Giguere, Alexandra Reingold, Allyson Smead, Lilly Torrey, and Ibardo Zambrano. We also thank Lucas Chute, Olivia Duong, Matthew Isacson, Jessada Mahatthananchai, Sam Merwin, Claire Stawski, and Diane Yeung for their work on several fronts.

The catalog for *Molecules That Matter* has been brilliantly designed by Barbara Glauber of Heavy Meta Design, and Shelley Wilks Geehr, Julie Conners, and Patricia Wieland of the Chemical Heritage Foundation coordinated joint work on the catalog. We thank them and authors Mary Ellen Bowden, Neil Gussman, Cyrus Mody, and Audra Wolfe of CHF; as well as Hassan Lopez, Mary C. Lynn, and Vasantha Narasimhan of Skidmore; Robert Hargrove of Mercer University; Daniel R. Goldberg of Boehringer Ingelheim; and Jeffrey Meikle of the University of Texas, Austin, for their expertise and insights.

Phil Fraley Productions of Paterson, New Jersey, has played a crucial role in the exhibition, elegantly fabricating the large-scale, scientifically accurate molecular models that form the conceptual core of *Molecules That Matter*. Our special thanks to Phil Fraley himself and his coworker Tim Ragan and collaborator Greg Balthus for their excellent work.

Ray and I join in thanking Mel Schaffer of Trademark Plastics for his insights into the nature and history of the plastics industry in the United States and for making it possible for us to tour production facilities in Leominster and Lawrence, Massachusetts, the birthplace of a wide range of American plastics.

Sincere thanks go also to Joseph Labovsky, the last surviving member of Wallace Carothers's lab at DuPont, whose video interview in the exhibition provides a uniquely personal look at Carothers and the historic work he accomplished.

We thank all the artists in the show, many of whom worked closely to select specific pieces and discuss the nature of the project with us, including Bryan Crockett, Chrissy Conant, Robert Dawson, Kara Daving, Melissa Gwyn, Michael Oatman, Alexis Rockman, and Jean Shin. We also thank the lenders to the exhibition, listed separately in this publication. Without their understanding and support this unique exhibition would never have been possible.

Finally, at the Tang we recognize the efforts of Susi Kerr for her work with students on the online feature; Elizabeth Karp for arranging loans, shipping, and the *Molecules That Matter* tour; Chris Kobuskie, Torrance Fish, and the Tang's superb on-call installation crew for installing the show beautifully and strategizing efficient travel and crating; Patrick O'Rourke for his design and signage; Ginny Kollak for editing and curatorial assistance; Ginger Ertz for her work with our community and school audiences; Gayle King for overseeing budgets; and Ian Berry for his advice and consultation in so many areas. Finally, I wish to single out the stellar work of the project's dedicated curatorial assistant, Kristen Carbone, Skidmore 2003, without whom this exhibition surely could never have happened.

My thanks to all for creating a unique, challenging, illuminating, and daring look at how molecular chemistry affects all of our lives decisively, every day, in ways we seldom perceive or understand. Making the invisible visible, *Molecules That Matter* helps all its viewers understand how much their lives, and life in our time, owe to organic chemistry and the scientists who extended its depth and breadth across the span of the twentieth century. Thank you, Ray, and all of your colleagues!

JOHN S. WEBER
DAYTON DIRECTOR
THE FRANCES YOUNG TANG TEACHING MUSEUM
AND ART GALLERY AT SKIDMORE COLLEGE

LENDERS

Brooke Alexander Editions
New York City

Pulickel Ajayan's Carbon
Materials Group
Rennselear Polytechnic Institute

James Cohan Gallery
New York City

Contemporary Art Museum
St. Louis

Dittrick Medical History
Center
Case Western Reserve
University

Bob Ensign

Raymond Giguere

Paul Ha

B. G. and Tamar Hacker

Phil Haggerty

Hagley Museum and Archive
Wilmington, Delaware

Robert Hargrove

Higgins Armory Museum
Worcester, Massachusetts

Matthew Isacson

The Joy of Giving
Something, Inc.
New York City

Gayle Rettew King

Diane and Howard Kloss

Leo Koenig, Inc.
New York City

Los Angeles County
Museum of Art

New York State Department
of Environmental
Conservation
Albany

Pember Museum of Natural
History
Granville, New York

Bernard Possidente

Andrea Rosen Gallery
New York City

A.G. Rosen
United Yarn Projects

Jordan D. Schnitzer
Collection
Portland, Oregon

John and Karen Sciolino

Skidmore College Chemistry
Department
Saratoga Springs, New York

Speed Museum
Louisville, Kentucky

Joan M. Stenerson

Frederieke Taylor Gallery
New York City

James W. Tribley

PHOTO CREDITS

The publishers would like to thank the artists, copyright holders, and rights holders of all images and texts in this catalog for granting permission to reproduce their works. Every effort has been made to contact copyright holders prior to publication.

All images reproduced in this catalog are photographed by Art Evans, and copyright of the Frances Young Tang Teaching Museum and Art Gallery, except the following:

Page 21: © Bayer AG/Corporate History & Archive. Page 23: Courtesy of the artist and James Cohan Gallery, New York City. Pages 37 and 38: Courtesy of the artist. Page 45: Collection of Brooke Alexander Editions. Page 53: Courtesy of the Joy of Giving Something, Inc., New York City. Page 55: © Bettman/CORBIS. Page 58: © *Life* magazine. Page 59: Reprinted through the courtesy of the editors of *Time* magazine © 2008 Time Inc. Page 60: Hill, John W; Kolb, Doris K., *Chemistry for Changing Times*, 10th edition, © 2004, page 597. Reprinted with permission of Pearson Education, Inc., Upper Saddle River, NJ. Page 68: Photograph by John Weber. Page 71: Courtesy of the artist. Page 73: © *Life* magazine. Page 77: Courtesy of the artist and Andrea Rosen Gallery. Page 82: Courtesy of the artist. Page 84: Courtesy of Hagley Museum & Library. Page 85: Courtesy of Hagley Museum & Library. Page 90: © Bettman/CORBIS. Page 99: Courtesy of the artist and Leo Koenig, Inc. Page 101 and 103: Image courtesy Vasantha Narasimhan. Page 106: Photograph by Rosalind Franklin. Page 109: © A. Barrington Brown/Photo Researchers, Inc. Page 110: David Mack/Photo Researchers, Inc. Page 115: © Bettman/CORBIS. Page 116: © Bettman/CORBIS. Page 121: Courtesy Joseph Farris.

Page 124: Courtesy of A.G. Rosen. Page 135: Used with permission of the McGraw-Hill Companies. William P. Cunningham, Mary Ann Cunningham, and Barbara Woodworth Saigo, *Environmental Science: A Global Concern*. McGraw-Hill, 2006. Page 138: Courtesy of the artist. Page 145: Courtesy *The New Yorker* Collection, 2006. Drew Dernavich from cartoonbank.com. All rights reserved. Page 148: © Sinauer Associates. Courtesy Meyer and Quenzer: *Psychopharmacology: Drugs, the Brain and Behavior*; Figure A in box 3.1. Page 154: Source: *After the Prefrontal Cortex* by J. M. Fuster, 1989. Copyright 1989 Ravan Press. Reprinted by permission. Page 160: Photo courtesy Richard E. Smalley Collection, Chemical Heritage Foundation Collections. Page 160: © IBM. Page 161: Photo courtesy Richard E. Smalley Collection, Chemical Heritage Foundation Collections. Page 164: Courtesy of the Higgins Armory Museum, Worcester, Massachusetts. Page 167: Photo courtesy Richard E. Smalley Collection, Chemical Heritage Foundation Collections. Page 171: Kroto, H.W., A.W.Afflaf, and S.P. Balm (1991) "C_{60}: Buckminsterfullerene," *Chemical Reviews* 91 (1985), 1,213-1,235 © Jim Heath. Page 172: Photo courtesy Richard E. Smalley Collection, Chemical Heritage Foundation Collections.

This catalog accompanies the exhibition *Molecules That Matter* organized by the Frances Young Tang Museum and Art Gallery at Skidmore College and the Chemical Heritage Foundation.

Chemical Heritage Foundation
315 Chestnut Street
Philadelphia, PA 19106
www.chemheritage.org

The Frances Young Tang Teaching Museum and Art Gallery at Skidmore College
Skidmore College
815 North Broadway
Saratoga Springs, NY 12866
www.skidmore.edu/tang

Designed by Barbara Glauber, Hilary Greenbaum, and Erika Nishizato/Heavy Meta, NY
Printed in Italy by Graphicom

EXHIBITION DATES AND TOUR

Tang Teaching Museum and Art Gallery
Saratoga Springs, New York
September 8, 2007–April 13, 2008

Chemical Heritage Foundation
Philadelphia, Pennsylvania
August 15, 2008–January 30, 2009

The College of Wooster Art Museum
College of Wooster, Wooster, Ohio
March 24, 2009–May 10, 2009

Faulconer Gallery
Grinnell College, Grinnell, Iowa
September 25, 2009–December 13, 2009

WEB SITE

http://tang.skidmore.edu /pac/mtm/

The *Molecules That Matter* Web site complements the gallery exhibition and catalog, providing additional historical information and scientific background about each molecule.